LEARNING THROUGH ENQUIRY

Making sense of geography
in the key stage 3 classroom

LEARNING THROUGH ENQUIRY

Making sense of geography
in the key stage 3 classroom

Margaret Roberts

Geographical
Association

Author

Margaret Roberts is a Senior Lecturer in the School of Education, University of Sheffield.

ISBN 978-1-84377-095-4
First published 2003
Impression number 10 9 8 7 6 5 4 3
Year 2013 2012 2011

Published by the Geographical Association, 160 Solly Street, Sheffield S1 4BF. Tel: 0114 296 0088; Fax: 0114 296 7176; E-mail: info@geography.org.uk; Websites: www.geography.org.uk and www.geographyshop.org.uk. The Geographical Association is a registered charity: no 1135148.

The Publications Officer of the GA would be happy to hear from other potential authors who have ideas for geography books. You may contact the Officer via the GA at the address above.

Edited by Rose Pipes

Index by Liz Cook

Designed by Arkima Ltd, Leeds, UK

Printed and bound in China through Colorcraft Ltd., Hong Kong

Unless otherwise acknowledged all photographs were taken by Margaret Roberts

CONTENTS

The term 'geographical enquiry' was first used about 30 years ago and is now part of the current orthodoxy of geographical education. 'Geographical enquiry' is statutory in the geography national curriculum in England, it is included in geography GCSE and Advanced level examinations and it is inspected by Ofsted. We are expected to use an enquiry approach to teaching geography. But what does it mean?

This book sets out to explore the meaning of 'geographical enquiry' at key stage 3, the phase of education on which most of my research has been focused. I have chosen to exclude fieldwork from the book because, although the process of enquiry is highly relevant to fieldwork, much has been written about it already. In contrast, relatively little has been written about geographical enquiry in the classroom.

*'The first key to wisdom is constant questioning …
By doubting we are led to enquiry, and by enquiry we discern the truth'*
(Peter Abelard, AD 1079-1142).

The meaning of enquiry (or 'inquiry' as it is sometimes spelt) has developed over time as it has been applied to different academic disciplines and professions at a range of levels. So there is nothing particularly geographical about 'enquiry'; it is the context in which it is applied that makes it geographical. My research has shown that what geography teachers understand by enquiry varies considerably, according to the particular contexts in which they have encountered it and from the ways they have incorporated it into their own practice.

Similarly, my own understanding of enquiry has been influenced by different strands of my career. Soon after I started teaching geography I became involved in the 'Language across the curriculum' movement, with its emphasis on students' use of language to make sense of what they were learning. Then I worked with others on the Nuffield Foundation Resources for Learning Project, exploring ways in which students, through the use of resources, could have more control over their own learning. At Countesthorpe College in Leicestershire, I worked alongside teachers who were interested in ways in which students constructed knowledge and ways in which teachers could support this process. My understanding of enquiry has also been informed by my research into the geography national curriculum, into enquiry work at key stage 3 and into the use of the internet. In addition, I have been able to see through my work, first as a teacher and then with PGCE students, the capacity of school students of all levels of achievement to become curious, critical and aware through their active involvement in the learning process.

So, what I understand as geographical enquiry has been shaped by these experiences, which have helped me develop convictions about how students learn and beliefs about what constitutes a worthwhile geographical education. I see geographical enquiry as an active process through which learners construct knowledge about the world. In order to learn, students need to make connections between what they already know and new information and new ways of seeing things. I think that they do this through the process of enquiry.

If geography is to be worth learning, then geographical enquiry should help students make sense of the world they live in and to make sense of what they hear, see and read about the world in their everyday lives; geography should help them make sense of their different personal worlds. In my opinion, geographical enquiry should be focused on real issues, on places and spaces that mean something to students and on real data of the kind that students are likely to encounter in the world outside the classroom. The emphasis of this book is implicit in its title. It is about learning geography, making sense of the world, through the process of enquiry. It is not about developing enquiry skills for their own sake, however transferable they might be.

Let me now turn to the structure of the book. Part One is concerned with general issues related to enquiry, literacy and numeracy. It explores how and why geographical enquiry has been incorporated into the geography curriculum and presents what I consider to be its essential elements (Chapters 1-3). I introduce a framework for learning through enquiry and apply this to examples throughout the book.

When I started to write I planned one chapter on language and learning. The fact that this has expanded to four chapters (Chapters 4-7) emphasises the importance I attach to the role of language in learning through enquiry. Almost everything students do in the geography classroom involves the use of language; they listen and talk; they read and they write. Even when they are engaged in the use of maps, pictures, graphs and diagrams, students use language in order to make sense of them. If we are to help students to learn through enquiry, then we need to understand the sense that they are making of things. To do this we need to get inside their minds, we need to pay attention to what they are saying and what they are writing and to their struggles to make sense. Attention to language is inevitably part of geographical enquiry. As much of the data used in geographical enquiry demands at least some understanding of number, Chapter 8 is devoted to numeracy.

In Part Two I develop ideas for different kinds of enquiry, focused on different types of questions. The examples included here (as in Part One) are intended to be illustrative, and are not presented as recipes to be followed exactly. The ingredients of a classroom situation are always uncertain, often because of the unpredictability of the key classroom ingredient, namely the students. Although many of the examples were developed in contexts with a relatively strong degree of teacher control, I would be very pleased if teachers developed them in such a way as to give students greater choice and greater control over their own learning.

This book is written primarily for teachers of geography, but it is hoped that it will be of interest to policy makers and to those carrying out research into geographical education. Although I have written the book within the context of the geography national curriculum in England, the framework, ideas and examples presented here are sufficiently general to be applicable to geographical education in other countries.

This book is intended to support teachers as professionals. The activities should be experimented with, changed, developed, adapted and applied in different contexts. The frameworks are presented to stimulate thinking and curriculum development rather to provide straightjackets. This book is my contribution to the ongoing debate about the nature of geographical enquiry and how it can be used to enable students to make sense of the world as they experience it.

Margaret Roberts

December 2002

ACKNOWLEDGEMENTS

This book would not have been possible without the period of study leave I was granted by the University of Sheffield. Many thanks to Julia Davies, Chris Winter, Ann Whorton and Chris Ash who covered my teaching and administrative responsibilities during my absence.

Although the book has been written during and since the period of study leave, the ideas in it have been developing throughout my career. I was fortunate, in the first 14 years of my teaching career, to work alongside teachers who were truly inspirational. Three in particular helped to transform my view of teaching and learning and continue to inspire me: Douglas Barnes (who was Head of English at Minchenden School Southgate, London where I started my teaching career); Michael Armstrong (with whom I worked both on the Nuffield Foundation Resources for Learning Project and at Countesthorpe College, Leicestershire); and Pat D'Arcy (who was also at Countesthorpe College).

I would also like to acknowledge my debt to the research and ideas of others. Francis Slater's *Learning through Geography* (Heinemann, 1982), together with her many other publications, was influential not only on the title of this book but also on my thinking about enquiry and significance of values in geographical education. I would like to thank her for the encouragement she gave me to pursue my interests in language and learning, and for the support she gave me throughout my academic career. I would also like to thank Eleanor Rawling, with whom I have discussed issues related to enquiry on many occasions, and Chris Spencer who introduced me to the concept of 'affective mapping'. Perhaps the biggest thank you of all should go to the geography PGCE students whom I have encountered during my time at the University of Sheffield. I have learnt a lot from working, thinking and reflecting with them on their classroom experiences. Special thanks are due to teachers and PGCE students who have willingly allowed me to use examples from their classrooms (see list below), particularly Steven Wilson who developed the idea of the 'big project' enquiry described in Chapter 16, and those who patiently obtained parental permission for all the photographs of students.

Although I produced the manuscript, the staff at the Geographical Association have transformed it into a publication. I would particularly like to thank Diane Wright who has worked patiently and carefully on the manuscript.

Finally, I would like to thank my three children – Rebecca, Elizabeth and Sam. I have learnt so much about teaching, learning and the process of schooling from them and their friends. They have made me see things differently, which is what I think learning is about. I appreciate the encouragement they have given me to write this book.

Examples contributed by: Jenny Allen, All Saints R.C. High School, Sheffield; Rachel Atherton, Maiden Earley School, Reading; Lucy Coddington, Tupton Hall School, Chesterfield; Richard Davies, Brentwood County High School, Essex; Jane Ferretti, King Edward VII School, Sheffield; Nikki Flanagan, The Royal Latin School, Buckingham; Chris Holt, Wright Robinson Sports College, Gorton, Manchester; Giles Hopkirk, Danum School, Doncaster; Dean Jones, Wickersley School and Sports College, Rotherham; Carole Luker, King Edward VII School, Sheffield; Anna Mercy, Hanson School, Bradford; Alison Munk, Wickersley School and Sports College, Rotherham; Chris Pearson, Myers Grove School, Sheffield; Jeanette Shipley, Tupton Hall School, Chesterfield; Chris Wainman, The Bolton School, Bolton; Carol Webb, Wollaston School, Wellingborough; Kim Wilson, King Edward VII School, Sheffield; Steven Wilson, Great Baddow High School, Chelmsford, Essex; Caroline Young, Nobel School, Stevenage, Hertfordshire.

And the following PGCE and former PGCE students: Jennifer Bird, Stephen Brady, Graham Fairclough, Rachel Gregory, Anna Mercy, Chris Pearson, Helen Prescott, Charlotte Wade, Susan Wells and Caroline Young.

PART ONE
Enquiry, Literacy and Numeracy

We shall not cease
from exploration
And the end of all
our exploring
Will be to arrive
where we started
And know the place
for the first time
(T.S. Eliot, *Four Quartets*)

*'"When I use a word," said
Humpty Dumpty, "it means
just what I choose it to mean
– neither more nor less"'*

(Carroll, 1998, p. 190).

INTRODUCTION

In 1996 I interviewed eleven teachers as part of a small research study (see 'Research into practice', pages 15-16). They taught in good geography departments in comprehensive schools in four different local education authorities. The schools were chosen for the variety of the GCSE specifications they used and of their catchment areas. One of the questions was: 'What do you understand by the term geographical enquiry?'. The replies were interesting, revealing the fact that although the term 'enquiry' is usually used as if everyone knows exactly what it means, this is clearly not the case. Six of the teachers interviewed gave six very different responses, each emphasising different aspects of enquiry (Figure 4, page 16). In the years since I carried out the research, the word 'enquiry' seems to have been used more and more frequently, yet I imagine that different understandings still exist.

This chapter explores the origins of our understandings of 'geographical enquiry', starting with an examination of the 'official' meaning of 'enquiry' in the geography national curriculum. I then refer to my own research on enquiry in the national curriculum and raise some issues related to policy and practice. Next, I examine curriculum development projects that have been influential in developing what enquiry means in practice. Finally, I outline how enquiry is presented in other subjects of the national curriculum.

Figure 1: Extract from the GNC programme of study for key stage 3. Source: DfEE, 1999a, p.22.

Knowledge, skills and understanding

Teaching should ensure that geographical enquiry and skills are used when developing knowledge and understanding of places, patterns and processes, and environmental change and sustainable development.

Geographical enquiry and skills

1   In undertaking geographical enquiry, pupils should be taught to:

a) Ask geographical questions *(for example, 'How and why is this landscape changing?', 'What is the impact of the changes?', 'What do I think about them?')* and to identify issues

b) Suggest appropriate sequences of investigation *(for example, gathering views and factual evidence about a local issue and using them to reach a conclusion)*

c) Collect, record and present evidence *(for example, statistical information about countries, data about river channel characteristics)*

d) Analyse and evaluate evidence and draw and justify conclusions *(for example, analysing statistical data, maps and graphs, evaluating publicity leaflets that give different views about a planning issue)*

e) Appreciate how people's values and attitudes *(for example, about overseas aid)*, including their own, affect contemporary social, environmental, economic and political issues, and to clarify and develop their own values and attitudes about such issues

f) Communicate in ways appropriate to the task and audience *(for example, by using desktop publishing to produce a leaflet, drawing an annotated sketch map, producing persuasive or discursive writing about a place)*

ENQUIRY IN GNC 2000

Geographical enquiry has been given a prominent place in the geography national curriculum (GNC) for England (DfEE, 1999a). In the programme of study for geography, at every key stage, the first sentence refers to 'geographical enquiry' and the whole of the first paragraph is devoted to it (Figure 1). In the GNC attainment target, enquiry skills are included in every level description (Figure 2). Clearly, enquiry is important. But what does it mean? The wording of the statutory orders suggests that enquiry has two main characteristics:

1. Enquiry is an integral part of teaching and learning. The first sentence of the programme of study indicates that enquiry and skills are not taught separately, but that they are used 'when developing knowledge and understanding of places, patterns and processes'. The attainment target endorses this integral view of enquiry; enquiry skills are assessed alongside the assessment of knowledge and understanding in determining levels of achievement.

Level	Extract from each of the level descriptions
	Figure 2: Extracts from the GNC attainment targets. Source: DfEE, 1999a, p. 43.
1	[Students] use resources that are given to them, and their own observations, to ask and respond to questions about places and environments.
2	They select information using resources that are given to them. They use this information and their own observations to help them ask and respond to questions about places and environments. They begin to use appropriate geographical vocabulary.
3	They use skills and sources of evidence to respond to a range of geographical questions, and begin to use appropriate vocabulary to communicate their findings.
4	Drawing on their knowledge and understanding they suggest suitable geographical questions, and use a range of geographical skills from the key stage 2 or 3 programme of study [PoS] to help them investigate places and environments. They use primary and secondary sources of evidence in their investigations and communicate their findings using appropriate vocabulary.
5	They begin to suggest relevant geographical questions and issues. Drawing on their knowledge and understanding, they select and use appropriate skills and ways of presenting information from the key stage 2 or 3 programme of study to help them investigate places and environments. They select information and sources of evidence, suggest plausible conclusions to their investigations and present their findings both graphically and in writing.
6	Drawing on their knowledge and understanding, they suggest relevant geographical questions and issues and appropriate sequences of investigation. They select a range of skills and sources of evidence from the key stage 3 programme of study and use them effectively in their investigations. They present their findings in a coherent way and reach conclusions that are consistent with the evidence.
7	With growing independence, they draw on their knowledge and understanding to identify geographical questions and issues and establish their own sequence of investigation. They select and use accurately a wide range of skills from the key stage 3 programme of study. They evaluate critically sources of evidence, present well-argued summaries of their investigations and begin to reach substantiated conclusions.
8	Drawing on their knowledge and understanding, they show independence in identifying appropriate geographical questions and issues, and in using an effective sequence of investigation. They select a wide range of skills from the key stage 3 programme of study and use them effectively and accurately. They evaluate critically sources of evidence before using them in their investigations. They present full and coherently argued summaries of their investigations and reach substantiated conclusions.
Exceptional performance	They draw selectively on geographical ideas and theories and use accurately a wide range of appropriate skills and sources of evidence from the key stage 3 programme of study. They carry out geographical investigations independently at different scales. They evaluate critically sources of evidence and present coherent arguments and effective, accurate and well-substantiated conclusions. They evaluate their work by suggesting improvements in approach and further lines of enquiry.

2. Enquiry is presented as something that students do themselves. It is students who have to develop the skills of: asking questions; selecting sources of evidence; analysing and evaluating data; and reaching conclusions. It is students who have to demonstrate these skills when they are assessed. The emphasis of both the programme of study and the attainment target is on student involvement and activity.

An examination of the level descriptions reveals a model of how students' involvement and skills in enquiry should progress. Progression is characterised by increasing:

• complexity of the context of enquiry
• involvement of students in the planning of enquiries (suggesting questions and issues to be investigated, planning sequences of investigation)
• involvement of students in selection (of sources, skills, and ways of presenting data)
• ability to use a range of skills (presenting and analysing data, reaching conclusions)
• ability to be critical in evaluating sources and evidence.

What the GNC presents about enquiry conjures up a picture of key stage 3 geography classrooms in which students are actively involved in investigating every theme and place and in which they are taking increasing responsibility for their own learning. This is not how it is in most schools. This is not because the requirements of the GNC are being deliberately ignored. It is partly because of the nature of national curriculum policy documents and partly because teachers have their own understandings of enquiry which they bring to their own readings of the GNC document.

RESEARCH INTO POLICY

My research into policy construction suggests that national curriculum documents are inevitably limited in the messages they convey. Although my research was carried out in relation to the 1995 GNC (see 'Research into national curriculum policy', pages 13-14), its findings are relevant to this chapter. The messages conveyed by national curriculum documents are limited in four ways: by what is allowed by the legislation; by consensus; by what is perceived to be politically acceptable; and by restrictions of space.

1. The Education Reform Act of 1988 allowed for legislation on what should be taught but not on how subjects should be taught. What enquiry might mean in terms of classroom practice could not be set out in statutory orders.

2. The 1991 GNC contained only token reference to enquiry, partly because of lack of consensus about its importance within the Geography Working Group. In 1995 there was apparent consensus in the Advisory Group about enquiry, but the word 'investigation' was used instead of 'enquiry'. It was not until the 1999 GNC that enquiry was given a prominent place in the GNC (Figure 3).

Figure 3: Enquiry in the geography national curriculum.
Source: Rawling, 2001, p. 81.

1991 requirements	1995 requirements	1999 requirements
Geographical enquiry is mentioned at the beginning of each PoS, but not explained or integrated into statements of attainment or PoS	Geographical enquiry process is included as paragraph 2 of each PoS and as a section in each level description, but is not named.	Geographical enquiry is clearly outlined as one of four key aspects of geography and explicitly described for each PoS. It is linked with skills and integrated with content.

Research into national curriculum policy

In 1996, as part of my research into the national curriculum, I interviewed six members of the Advisory Group involved in the construction of the 1995 GNC. My aim was to investigate how and why 'enquiry' had been included in the document.

Context of the research

Following widespread dissatisfaction with the 1991 national curriculum generally, a committee was set up, chaired by Ron Dearing, to review the national curriculum and provide guidelines for its revision. The Dearing recommendations (SCAA, 1994) required a slimming down of the content, no new material, and a reduction in the number of attainment targets.

The starting point for revision in all subjects therefore was the 1991 national curriculum. For geography this was not the ideal starting point for constructing a curriculum in which geographical enquiry might have a significant place. In GNC 1991 the phrase 'enquiry approach' had been used but with no details of what it meant and there had been no requirement to assess any enquiry skills. Because of the structural problems of GNC 1991, the Advisory Group was allowed some opportunity to restructure the curriculum and this gave them some scope for including enquiry.

Research findings

The 1994-95 Advisory Group was strongly influenced by members who felt that geographical enquiry should have a significant place in the GNC. Five members of the group had been involved, for long formative periods of their careers, with one or more of the Schools Council geography projects and this had influenced their views on enquiry. Some interviewees were determined that, in spite of the constraints of the Dearing guidelines and of the 1991 Orders (DES, 1991), enquiry would be included: 'it was a case of trying to ensure that it was in there'.

There were four issues related to enquiry facing the group:

- dealing with the word 'enquiry';
- suggesting what enquiry meant in the classroom without dictating methods;
- ensuring that enquiry was seen as integral;
- determining some notion of progression.

One of the ways of including enquiry was, paradoxically, by not using the word 'enquiry' at all in the 1995 Orders (DfE, 1995), but by using the word 'investigate' in its place. I was given various explanations of this:

- An enquiry approach suggested a particular way of teaching and this was beyond the scope of the national curriculum legislation.
- The word 'investigation' was used in science, sounded more 'hard-nosed', had more credibility, and was therefore more acceptable. Its use had not given rise to 'worries about subversive methodology'.
- Enquiry had become a loaded word, 'some people respond against it straight away and other people are very pro it'.
- There was 'confusion amongst teachers' about what enquiry meant, particularly among non-specialists.
- The word 'enquiry' would not be approved by those higher up in SCAA (School Curriculum and Assessment Authority, now incorporated into the Qualifications and Curriculum Authority (QCA)).
- 'Enquiry' was associated with controversial issues.

Although members of the Advisory Group associated 'enquiry' with particular ways of teaching, the national curriculum legislation did not allow statutory orders to state *how* subjects should be taught, only *what* should be taught. 'Enquiry' or 'investigation' had to be included in a way that did not dictate teaching styles. However, the list of investigative skills did carry implicit messages about teaching style. For example, one requirement was that 'pupils should be given opportunities to identify geographical questions'. One member of the group said, 'it is difficult to envisage a teacher teaching that from the blackboard'.

Most members of the group understood enquiry as something integral to everything studied:

'I would like to see it emphasised that this [enquiry] isn't just a project, this is the investigative approach to the whole of the curriculum.'

'It was obvious within the whole of the Advisory Group that enquiry was to lead the whole programme of study and all the elements. Everything would be integrated. We all felt that we wanted to see each of the themes developed through the setting up of an enquiry.'

The way the group tried to indicate that an investigative approach should be used for all study was in the use of the phrases 'in investigating' and 'in studying' to introduce each of the themes, e.g. 'In investigating how environments change, pupils should be taught ...'. I was told that there was no significance in the fact that 'investigating' was used for some themes and 'studying' for others. Both words were

Research into national curriculum policy ... continued

intended to convey the message that investigative work was not something separate, but something to be applied to all themes. One interviewee, however, did not understand the phrases in this way and did not think that 'enquiry' should be used for everything.

'I think variety is important. I think an investigative approach is fine in certain aspects but I think there's also a place for other ways of doing the job. It shouldn't be the only way you teach things.'

Members of the group thought that an essential feature of the enquiry approach was the use of key questions, and exemplar questions were included in the first paragraph of each key stage. For example, at key stage 1, the exemplar questions were: 'What/where is it?, What is it like? and How did it get like this?' (DfE, 1995, p. 2). Progression was marked by the inclusion of additional questions at key stage 2, 'How and why is it changing?', and key stage 3, 'What are the implications?' (DfE, 1995, pp. 2 and 10). Some members of the group argued for two attainment targets for geography, one based on knowledge and understanding and one based on skills, including enquiry skills. It was felt by some that this would give the process of geographical enquiry more importance in the curriculum and would prevent it being overlooked in a general all-inclusive paragraph. Others argued for a single unified attainment target. The single attainment target for geography was chosen partly because of Dearing's wish to simplify the assessment procedure, partly because the GNC was expected to be similar to the history national curriculum in structure, and partly because of the weight of arguments within the group for each proposal.

Note: This research was presented at the British Educational Research Association conference in Bath in 1997.

3. What is considered to be politically acceptable depends on the dominant ideologies of those with power to influence political decisions. The 1988 Education Reform Act, which included legislation for the introduction of the first national curriculum for England and Wales, was influenced by the New Right. The New Right was critical of the more progressive ideologies promoted by the Schools Council projects which in geography had dominated curriculum development and the introduction of enquiry approaches in the 1970s and 1980s. Rawling argued that for the New Right 'geographical enquiry represented the previous era and seemed to represent all that was wrong with education – a lack of rigour, emphasis on woolly attitudes and general skills, and a lack of discipline and teacher control' (2001, p. 39). The New Right wanted to restore the traditional curriculum with its emphasis on subject knowledge and the skills associated with particular subjects. By the time the 1999 GNC curriculum was constructed, there had been a change in government, a change in educational priorities and a re-emergence of some progressive ideas.

4. Those who write national curriculum policy documents are inevitably limited by space. Ever since the production of the sets of ring binders containing the first national curriculum in the early 1990s there has been pressure to streamline the statutory orders. Although enquiry is prominent in GNC 2000, what is written on it amounts to no more than two sides of A4.

The implication of these limitations is that policy makers do not convey meaning as clearly as they would like. What is written is the result of processes of contestation, consensus, and negotiation (Ball and Bowe, 1992). The final product is a compromise, conveying clues to meaning which those advising on policy hope will be picked up. As one of those contributing to the construction of GNC 1996 said, 'We knew we were waving a flag. When anybody in geography read that they would say, "Oh a flag has been waved here for the enquiry method"'.

RESEARCH INTO PRACTICE

My interviews with teachers (see 'Research into practice', below) in 1996 suggested that they did not see 'a flag' waving for the 'enquiry method' when they read the new 1995 GNC documents. Teachers were busy with other curriculum changes and did not read GNC 1995 with close attention to every word. For example, none of those interviewed had noticed the phrases 'in investigating' and 'in studying', so carefully inserted by the policy makers before each theme. The teachers I interviewed had focused their attention on the content of the programme of study; they wanted to know the implications of the changes for their existing schemes of work which were structured on content.

Research into practice

Context

In 1996 I carried out a focused piece of research for SCAA 'to investigate how teachers in a range of schools are seeking to integrate enquiry questions and skills into geography schemes of work so as to encourage progression in this aspect of geography within and between key stages'. I collected data from geography departments in six comprehensive schools in four different local education authorities, chosen for variety in terms of examination syllabuses used at GCSE and A-level. The research was not intended to produce generalisations, but was designed to explore the variety of understandings of enquiry and to increase understanding of the way experienced practitioners work with national curriculum documents in practice. In each school I interviewed teachers and collected examples of schemes of work, assessment items and students' written work.

Research findings: understandings of 'geographical enquiry'

The interviews revealed different understandings of geographical enquiry (Figure 4). Although enquiry was commonly associated with fieldwork, independent learning and a sequence of investigation, each teacher attached different degrees of importance to these aspects and this had an impact on classroom practice. Where association of enquiry with fieldwork was very strong there was little enquiry work in the classroom.

There were other differences of emphasis in how the teachers conceptualised geographical enquiry. One department emphasised hypothesis testing and the scientific method, while another department put more emphasis on qualitative data, values and attitudes. The teachers were influenced in their thinking by their own personal biographies and by ways in which they had encountered enquiry in their own lives as students and teachers.

Research findings: geographical enquiry at key stage 3 in practice

These different understandings of enquiry influenced practice in schools, which varied considerably. The research findings related to practice were summarised for SCAA as follows:

1. Teachers distinguished between work that was enquiry based and work that was not.

2. The extent of enquiry-based work varied considerably between schools and to some extent within departments.

3. Schools varied in whether and how they incorporated geographical enquiry into assessment procedures.

4. The nature of geographical enquiry varied between schools and could be categorised according to which aspects of enquiry were emphasised. The categories identified were: field work; questionnaire surveys; library skills; structured flexible learning; decision making and problem solving; and coursework assessment tasks.

5. The nature of geographical enquiry varied according to how teachers worked with students in the classroom. The meaning of enquiry was constructed in the classroom through teacher and student interactions.

6. None of the case study schools planned for progression in geographical enquiry. The progression that existed arose from the way teachers intuitively varied their teaching between year groups and individuals.

7. There was no planned progression between key stages 2 and 3. Teachers in the six case study departments found it difficult to build on what students had done at key stage 2 because of the large number of feeder schools, with considerable differences in the geographical education they offered.

8. The contexts in which some teachers worked made it difficult for them to develop geographical enquiry as they would have liked.

9. Teachers tended to conceptualise the curriculum in terms of content and tended to ignore the enquiry requirements in the GNC.

10. Teachers were confused about how to apply the level descriptors and were unclear about the contribution of the enquiry strand in determining a best-fit judgement.

The findings of the study have been published in *Teaching Geography* (Roberts, 1998a) and were developed further to produce a discussion paper (QCA, 1998).

Figure 4: Different teachers' understandings of geographical enquiry.

Selected extracts from six research interviews (see 'Research into practice', page 15):

'We look upon enquiry as being a situation where students, preferably outside school, have an opportunity for independent learning based on a set of problems or questions or hypotheses, for which they can find the information and formulate answers. They have to set a hypothesis, because otherwise I don't see how they can go through the whole enquiry process.'

'Instead of somebody standing at the front telling you the information you try and find out yourself through other means.'

'Enquiry is when you formulate what you are going to look at and you find your own evidence. It is where people decide their own learning pathways, basically, and the nearer you get to handing the whole learning process over to the kids, the nearer you are to true enquiry-based learning.'

'It is enquiry-based when they go out, and they are doing a bit on their own … then they go home and do some follow-up work.'

'I see enquiry as a form of problem solving where they have to go through a route of asking questions, how are they going to find it out?, what is their response to it? I see it as a wide process but using a variety of evidence. They have to search out what is relevant, make connections. We don't do much number crunching. For most of the enquiry, values and attitudes come into it.'

'Supported self-study, where you have got a job, you have the resources and the students make their way through to a particular finished product. It is basically the route followed through in the MEG syllabus B: identify the problem, identify a method, identify resources, do something, appraise it and re-plan.'

How the teachers read the GNC was influenced not only by their focus on content but also by the understandings they brought to their reading. They had very different understandings of 'enquiry' (Figure 4) influenced by their own personal biographies. The extent to which they equated enquiry with fieldwork or with quantification was influenced by a range of factors such as the type of geography they had studied as students, the people they had encountered in their professional careers, the GCSE syllabuses they used and whether they had been involved in any of the Schools Council's geography projects.

The meaning they attached to the term 'investigation' in the programme of study influenced how enquiry was developed in their departments and what actually took place in the classroom. There were examples of good enquiry work taking place in each of the departments in the research project, but the meaning of enquiry was different. In one department enquiry was mainly associated with fieldwork and in another it was associated with student independence. In one department enquiry dealt mainly with quantitative data, in another mainly with qualitative data. In all of the departments studied, enquiry was considered as an occasional activity, not as an approach to learning everything prescribed in the programmes of study.

THE LEGACY OF THE SCHOOLS COUNCIL GEOGRAPHY PROJECTS

The influence of the projects

The term 'enquiry' started to be used in geographical education in relation to the curriculum development work of the Schools Council geography projects in the 1970s and 1980s. Each of these geography projects became associated with the term 'enquiry'. The Geography for the Young School Leaver Project referred to 'areas of enquiry'; the 14-18 Geography Project referred to 'inquiry skills, strategies and processes' (with enquiry spelt with an 'i'); the 16-19 Geography Project referred to 'enquiry-based teaching and learning'.

Each of these projects has been highly influential in developing present understandings of geographical enquiry, through teachers involved in the projects, through the examination system and through textbooks. The teachers involved in the research and pilot phases of the projects developed their understanding of enquiry by applying the projects' frameworks to their schemes of work, their lesson planning and their classroom practice. They developed their own meanings of enquiry in the classroom as they adapted the project frameworks within their own school contexts (Parsons, 1987; Dalton, 1988).

Many of the teachers involved in the early stages of the projects have had influential positions in geographical education in their schools, in local education authorities and nationally.

Each of the three projects was developed into an examination syllabus and each of these has been further developed into GCSE specifications used at present. Ways of thinking about geographical education embodied in these specifications have become deeply ingrained and influence how teachers plan the curriculum and think about enquiry, not only for these examinations, but also at key stage 3 (Roberts, 1996, 1998b). Many textbooks have been written specifically for the examinations based on the Schools Council projects and these textbooks in their turn influenced ideas about enquiry.

Ideas about enquiry introduced by the projects

There were several common characteristics of enquiry in the Schools Council geography projects.

1. Enquiry was seen as a classroom activity as well as a fieldwork activity.

2. Enquiry was seen both as an approach to learning and as a basis for an individual study.

3. Enquiry was seen as an integral part of a geography course; it was not an end in itself. It was the process by which geographical content and concepts were studied.

4. Enquiry involved a shift in teaching style from one in which information and ideas were transmitted by a teacher to one in which students were engaged in activities.

5. Enquiry involved the use of a wide range of resources, used as evidence.

6. Enquiry developed a wide range of skills.

In addition to these common characteristics, each project developed particular ways of thinking about enquiry.

The Schools Council GYSL Project

This project (see panel below), initiated in 1970, set out to be relevant to the interests of those leaving school at the end of the compulsory period of schooling. The project put emphasis on relevance, the understanding of ideas, and innovative ways of working. It developed a tightly controlled view of what enquiry meant. There were detailed predetermined objectives. The resources and activities were chosen by the project or by the teachers in order that students might achieve these objectives. The only scope for more open-ended enquiry was in students' individual studies.

The Schools Council GYSL Project

Background

The Schools Council 'Geography for the Young School Leaver' (Avery Hill) Project was set up in 1970 to develop a course for 14-16 year olds of 'below average to average ability'. The project set out to provide a curriculum that was to be 'pupil-centred in the sense of being related to areas of life and experience which they see as relevant to them – relevant in terms of the areas of enquiry studied and relevant in terms of the learning styles which equip them as young adults in society' (Schools Council, 1980, p. 12). It was concerned with all aspects of students' development and defined its learning objectives in terms of ideas, skills and values. The emphasis was on the understanding of key ideas and these provided the focal point for the selection of content.

Teaching style

The project wanted to encourage a teaching style in which students had an active role. The following guidelines were provided in the Teachers' Guide:

'Each unit of the theme contains a wide range of resources – discussion sheets, slides, newspaper extracts, photo-sheets, maps, statistics etc. These are intended to be used flexibly by the teacher, to enable him [sic] to design learning experiences which will lead to the achievement of stated objectives.

'The resources provide the basis for pupil centred activities. By seeking answers to problems, individual thinking is encouraged and this replaces memorisation as a dominant classroom activity. It should be possible to create learning experiences which will enable the pupil, whatever his ability or level of motivation, to test evidence,

to interpret, to use his own judgement, to be aware of his own and other people's attitudes and to be imaginatively involved in creative situations. He may be working individually or in one of a variety of suggested group situations.

'The structure of the content of the theme ensures that while discovery methods encourage individual approaches there is continuity and sequence in the overall pattern. The established pattern of observation, interpretation and generalisation is extended to allow the pupil to test out generalisations in other situations' (Schools Council, 1980, p. 15).

Example GYSL objectives from a unit of work on 'Movement in the city'

Part 1: Individual movements within the community

Objectives

Ideas

- Movement to work, services, shopping and leisure varies from one individual to another
- In some neighbourhoods the range of provision necessitates less movement than in others

Skills

- Map interpretation, location and distance-measurement

Values and attitudes

- A concern with the degree of adequacy of, and accessibility to, essential community needs.'

The Schools Council Geography 14-18 Project

This project (see panel opposite) was set up in 1970 to develop a curriculum for 'more able' students. It put emphasis on students using enquiry processes 'similar to those which geographers themselves follow': students were expected to think like geographers. The project recognised different 'traditions of inquiry' in higher education and encouraged the use of different enquiry processes. The project defined broad aims rather than detailed objectives and allowed for differences in outcomes, unexpected outcomes and student choice. The starting points for constructing units of work were 'basic questions', 'fundamental concepts' and enquiry processes.

Background

The Schools Council Geography 14-18 Project (the Bristol Project) was set up in 1970, 'to initiate a programme of curriculum development in geography for more able pupils which would offer them an intellectually exacting study and contribute more substantially to their general education' (Tolley and Reynolds, 1977, p. 1).

There was, in 1970, dissatisfaction with what school geography offered to 'more able' students over the age of 14. School geography took little account of the exciting changes taking place in academic geography at that time and there was too much emphasis in examinations on factual recall. The Geography 14-18 Project aimed to develop a curriculum that would be 'intellectually challenging' and that would be relevant to 'real world issues'. This involved changes in both the content of what was taught and also in teaching style. In terms of content, the project wanted to reflect the new ideas of academic geography. In terms of teaching style, the project wanted to take greater account of students' interests, to encourage more classroom interaction and to get students involved in 'activities which involve them in processes of inquiry [sic] similar to those which geographers themselves follow when attempting to solve problems'.

The project team put considerable emphasis on the professional development of teachers through their involvement in the project and through their application of the project's guidelines for curriculum development in their own schools.

The role of geographical enquiry in the project

The concept of enquiry presented in the 14-18 project was underpinned by two central ideas. First, the project believed that the fundamental concepts of the subject and its enquiry processes were inextricably linked. Teachers needed to take concepts and enquiry processes into account in curriculum planning. Second, the project thought that students should use the same intellectual processes in the classroom as were being used by academic geographers: 'The pupils must go through processes of reflection and inquiry similar to those which gave rise to geography as an organised body of knowledge in the first place. In other words, they must learn to think geographically' (Tolley and Reynolds, 1977, p. 21).

Aims

'Enquiry' was an integral part of the broad aims of the project:

'Courses in geography at 14-18 should have:

a. a medium-term aim, that the pupil has sufficient grasp of the forms of inquiry and understanding in geography to appreciate their possibilities for further study

b. a long-term aim, that the understanding which the pupil has developed in the subject will modify his perceptions and sensibilities in ordinary life.'

It is interesting that the aims referred to 'forms' of enquiry rather than 'form' of enquiry. The project recognised the value of different traditions of enquiry in academic geography and did not want to limit courses to one tradition. The project emphasised the value of the course, not only as a foundation for further academic study, but as of value in 'ordinary life'.

By the end of the course students were expected to be able to use the following enquiry processes, 'in forms appropriate to the context:

- discern significant problems
- review relevant sources
- formulate appropriate generalisations
- collect evidence systematically
- appraise evidence against criteria, come to balanced conclusions, report findings honestly
- revise initial views
- anticipate the possible indirect effects of intervention in complex physical and ecological systems
- communicate ideas clearly and in ways appropriate to the context' (Tolley and Reynolds, 1977, pp. 19-20).

The project emphasised broad aims rather than detailed specific objectives: 'There is a danger in allowing one's teaching to be too centred on particular learning objectives. The teacher's role is less to transmit specific content than to exemplify the criteria of rational investigation and the arts of inquiry' (Tolley and Reynolds, 1977, p. 152). The project wanted to allow for individual and unexpected outcomes of learning and did not want to over-determine how students responded to resources and problem-solving activities.

The examination based on the 14-18 project

The 14-18 syllabus had three elements: the core units; coursework; and individual study.

1. The core units incorporated fundamental conceptual frameworks and basic strategies. Teachers could choose

Figure 5: The route for geographical enquiry. Source: Geography 16-19 Project, School Curriculum Development Committee, 1985.

FACTUAL ENQUIRY more objective data	ROUTE AND KEY QUESTIONS	VALUES ENQUIRY more subjective data
Achieve awareness of a question, issue or problem arising from the interaction of people with their environments.	Observation and perception What?	Achieve awareness that individuals and groups hold differing attitudes and values with regard to the question, issue or problem.
Outline and define the question, issue or problem. State hypotheses where appropriate. Decide on data and evidence to be collected. Collect and describe data and evidence.	Definition and description What? and Where?	List the values held or likely to be held by different individuals or groups with interest and/or involvement. Collect data on actions and statements of individuals/groups. Classify values into categories. Assess the actions likely to be linked with each category.
Organise and analyse data. Move towards providing answers and explanations. Attempt to accept, reject or modify hypotheses. Decide whether more or different data and evidence are required.	Analysis and explanation How? and Why?	Assess how far the values can be verified by evidence, i.e. to what extent are the values supported by facts? Attempt to recognise bias, prejudice, irrelevant data. Identify sources of values conflict.
Evaluate results of enquiry. Attempt to make predictions, to formulate generalisations and, if possible, to construct theories. Propose alternative courses of action and predict possible consequences.	Prediction and evaluation What might? What will? With what impact?	Attempt to identify the most powerful values positions. Consider future alternatives from these positions and recognise preferred decisions. Identify people/groups who could act and assess impacts/consequences.
Recognise the likely decision given the factual background and the values situation. Identify the probable environmental and spatial consequences.	Decision making What decision? With what impact?	Recognise the likely decision given the results of the values analysis and the factual background. Identify the probable reactions and responses of those who hold other viewpoints.

Personal evaluation and judgement
What do I think? Why?
Determine what values are important to oneself and so decide which values position one would support in this issue.
Identify which decision and what courses of action one could accept personally.
Assess their impact on the situation.
Consider how one would defend and justify this course of action.

Personal response
What next? What shall I do?
Decide whether as a result of this enquiry:
• to take action oneself or with others on this issue
• to help initiate action on this issue by contacting those in positions of power
• to take action to change aspects of one's personal lifestyle/actions which may affect future issues
• to take no immediate action, but to follow further enquiries in order to test out one's feelings

their places to exemplify the common ideas. Teachers selected a wide range of resources and devised challenging activities. There was an emphasis on the use of models, systems and spatial analysis. Units of work were structured around a series of questions. For example, the project suggested that the following questions would be suitable for a unit of work on transport: What is the problem? How does the system work? How can the system be managed? What values are involved? What trends must be anticipated?

2. Coursework could be partly through individual study and partly through teacher-planned enquiry.

3. Individual studies provided students with the 'opportunity for independent self-directed study and for gaining experience in the arts and strategies of inquiry in geography' (Tolley and Reynolds, 1977, p. 196).

The Schools Council 16-19 Project

This project (see panel) was set up in 1976 to redevelop the post-16 geography curriculum. The project put emphasis on an integrated people-environment view of the subject and on an integrated view of enquiry combining 'factual enquiry' and 'values enquiry'. The starting point for enquiry was 'necessarily a question, problem or issue'. The project developed a 'route for geographical enquiry', a sequential framework of questions and activities, which has been influential on subsequent thinking (Figure 5). Rawling considered that although all the projects explored enquiry approaches to learning, 'it was the 16-19 geography project that presented the most complete exposition' (2001, p. 38).

Schools Council 16-19 Geography Project

An enquiry-based approach to teaching and learning

The Schools Council 16-19 Geography Project, set up in 1976, was aimed at curriculum renewal for post-16 geography. The project envisaged that 'the enquiry-based approach to learning' could encompass a range of teaching methods and approaches, from activities such as problem solving, structured by the teacher, to more open-ended discovery. It believed that a learning situation had an 'orientation towards enquiry' if it satisfied Dewey's definition of reflective thought as 'active, persistent and careful consideration of any belief or supposed form of knowledge in the light of grounds that support it and the conclusions to which it tends' (1933, p. 6). According to the project, what distinguished enquiry-based learning from other approaches to learning was teacher encouragement to students 'to enquire actively into questions, issues and problems, rather than merely to accept passively the conclusions, research and opinions of others' (Naish *et al.*, 1987, p. 45).

Characteristics of the 16-19 enquiry-based approach

The project set out the characteristics of the enquiry-based approach to learning as *'an approach to learning which:*

- *identifies questions, issues and problems as the starting points for enquiry*
- *involves students as active participants in a sequence of meaningful learning through enquiry*
- *provides opportunities for the development of a wide range of skills and abilities (intellectual, social, practical and communication)*
- *presents opportunities for fieldwork and classroom work to be closely integrated*
- *provides possibilities for open-ended enquiries in which attitudes and values may be clarified and an open interchange of ideas and opinions can take place*
- *provides scope for an effective balance of both teacher-directed work and more independent student enquiry*
- *assists in the development of political literacy such that students gain understanding of the social environment and how to participate in it'* (Naish et al., 1987, p. 46).

Schools Council 16-19 Geography Project ... continued

The route for geographical enquiry

The Geography 16-19 Project developed, as an organising framework to guide curriculum planning, a 'route for geographical enquiry' (Figure 5). The 'route for enquiry' provided a sequence of questions and student activities, to be used in studying real issues, preferably topical, supported by data as evidence. The 'route for enquiry' incorporated several aspects of enquiry that were innovative in the 1970s and 1980s:

1. Instead of values being treated as something separate in enquiry work, often investigated after the investigation of the 'facts' of an issue, the route for enquiry encouraged the use of 'more objective data' and 'subjective data' throughout the investigation of an issue. The project's view was that values were integral to enquiry work and that 'explanation which ignores attitudes and values is likely to be arid and meaningless' (Naish *et al.*, 1987, p. 174).

2. The project increased the range of questions addressed by school geography. In addition to traditional enquiry questions (What? Where? How? and Why?), the 'route for enquiry' included questions such as: What might? What will? What decision? With what impact? What do I think?

3. The 'route for enquiry' presented a view of geographical education in which different aspects of study were closely integrated. The development of skills was integrated into the study of issues. The study of the physical environment was integrated with the study of human geography. The study of theory was related to context. The study of 'objective' data was related to the study of 'subjective' data.

Legacy and limitation of enquiry as developed by the Schools Council projects

Our present understanding of geographical enquiry owes a lot to how the meaning of enquiry was developed by the Schools Council projects and by the examinations based on their principles. The projects introduced new ways of thinking about teaching styles, they integrated the development of enquiry skills into the study of subject content and they broadened the range of skills developed and resources used.

In several ways, however, the projects presented a limited view of enquiry. The way the projects were developed in practice led to a teacher-centred type of enquiry work. Within the framework of the syllabuses, teachers tended to choose the content to be studied, the questions or key ideas that framed the enquiry, the resources and the activities. This meant that students had opportunities to develop some enquiry skills much more than others. They developed the skills of presenting, analysing and interpreting data and reaching conclusions. They rarely developed skills of asking questions, choosing content, searching for data, and selecting data, except in the context of individual studies.

The Schools Council projects were developed into examination courses, with their detailed predetermined assessment objectives. The specification of course content and detailed objectives limits the nature of enquiry work. There are problems in defining content and objectives and at the same time encouraging students to define questions and issues for themselves and to construct their own meanings from enquiry work. If objectives are tightly defined, then there is less scope for student choice, selection, and for unanticipated outcomes.

'ENQUIRY' IN GCSE SYLLABUSES AND SPECIFICATIONS

The examinations based on the Schools Council projects have promoted, through the way they are assessed, two ways of thinking about enquiry:

1. Enquiry is an approach to learning and this approach influences the way the syllabuses (and now specifications) have been assessed.

2. The individual investigative study has also been referred to as 'an enquiry'.

However, it is not only examinations based on Schools Council projects that have influenced understandings of enquiry. 'Enquiry' has formed part of all GCSE and A-level geography examinations for many years, but in the sense of 'a piece of investigative work to be tested separately, rather than as an overarching approach to learning underlying all assessment activities' (Rawling, 2001, p. 111). Syllabuses and specifications have often included diagrams of the process of enquiry, but presented only in relation to the individual study - the separate piece of enquiry work. The examinations teachers have used in their schools have had a significant influence on how they think about enquiry, how they understand what is required by the GNC (Roberts, 1998a) and how they develop enquiry in the classroom at key stage 3 (Roberts, 1995, 1998b).

'MAN: A COURSE OF STUDY'

Although the Schools Council geography projects developed meanings of enquiry for geography teachers, there had been earlier curriculum development projects developing enquiry methods of learning. One of the first projects to use a process model of curriculum planning was 'Man: A Course of Study' developed by the psychologist Jerome Bruner in 1966, for 10-12 year old social studies students in the USA (see panel below). Bruner expressed the aims of the project in terms of principles rather than objectives. The principles were intended to underpin all learning rather than represent endpoints of learning. The first principle was: 'To initiate and develop in young people a process of question posing'. Although Bruner devised the three overarching questions himself, he planned activities that would promote speculation and questioning and thinking about thinking. Bruner's ideas, although developed over 35 years ago, are relevant to the development of enquiry in the geography classroom today. His emphasis on fundamental concepts, processes of enquiry, questioning, problem solving, and reflection are remarkably similar to the concerns of the 'Teaching and Learning in the Foundation Subjects' strand of the National Strategy for Key Stage 3. 'Man: A Course of Study' was concerned with big ideas, the thinking skills involved in enquiry and metacognition.

'Man: A Course of Study'

Background

In 1966 the psychologist, Jerome Bruner, constructed a course of study based on 'the enquiry method' of learning. His aim was to construct a social science course for 10-12 year olds, a course that did far more than 'get something across' or 'merely impart information'. He was concerned with developing the intellectual powers of students. The course, 'Man: A Course of Study', consisted of an elaborate package of student materials together with teachers' books providing guidance on strategies. Three questions recur throughout the course:

- What is human about human beings?
- How did they get that way?
- How can they be made more so?

The course explored these questions through the study of five themes: tool making; language; social organisation; the

management of 'man's' prolonged childhood, and 'man's' urge to explain his world.

Ideals

Bruner wanted the course to achieve five ideals:

- 'to give our pupils respect for and confidence in the powers of their own mind;
- to extend that respect and confidence to their power to think about the human condition, man's plight, and his social life;
- to provide a set of workable models that make it simpler to analyse the nature of the social world in which we live and the condition in which man finds himself;
- to impart a sense of respect for the capacities and humanity of man as a species;
- to leave the student with a sense of the unfinished business of man's evolution' (Bruner, 1966, p. 101).

'Man: A Course of Study' ... continued

Aims

According to Bruner, the course had clear pedagogical aims:

- *'to initiate and develop in youngsters a process of question-posing (the inquiry method);*
- *to teach a research methodology where children can look for information to answer questions they have raised and use the framework developed in the course (e.g. the concept of life cycle) and apply it to new areas;*
- *to help youngsters develop the ability to use a variety of first-hand sources as evidence from which to develop hypotheses and draw conclusions;*
- *to conduct classroom discussions in which youngsters learn to listen to others as well as to express their own views;*
- *to legitimise the search: that is, to give sanction and support to open-ended discussions where definitive answers to many questions are not found;*
- *to encourage children to reflect on their own experiences;*
- *to create a new role for the teacher, in which he becomes a resource rather than an authority'* (Bruner, quoted in Stenhouse, 1975, pp. 38-9).

Techniques

Within the course Bruner developed four particular techniques to encourage student engagement:

1. *The use of contrast.* Children were given information about an unfamiliar topic or situation, e.g. 'man' versus higher primates, to enable them to appreciate what was too obvious and familiar to them.

2. *Stimulation and the use of informed guessing, hypothesis making and conjectural procedures.* The course included strategies to encourage 'informed guessing'.

3. *Participation.* The course included games and activities to promote the active involvement of children.

4. *Stimulating self-consciousness about thinking and its ways.* The course encouraged students to develop the skills of getting and using information to be conscious of their 'tools of thought', e.g. causal explanation, to have the experience of using theoretical models and to pause and review in order to recognise what they have learned.

ENQUIRY ACROSS THE CURRICULUM

Students develop enquiry skills in many other subjects of the national curriculum. Figure 6 shows an extract from *The National Curriculum: Handbook for secondary teachers in England (KS3&4) (DfEE, 1999b)*. Enquiry is included as one of five thinking skills which are to be applied to all subjects. What is rather odd about the list is that all the other skills listed - information processing, reasoning, creative thinking and evaluation - are all needed for different aspects of enquiry work. Enquiry skills are not a sub-set of thinking skills; enquiry includes them all.

Figure 7 (page 26) shows extracts from the key stage 3 programmes of study for six subjects, including geography. There are many similarities. All the subjects apart from ICT use the term 'enquiry'. Students are expected to consider issues in both geography and citizenship. They are expected to collect data and represent it in history, ICT, mathematics and science. In all subjects they are expected to reach conclusions. There are some differences in emphasis between subjects. For instance, critical evaluation is included in the programmes of study for ICT, mathematics and science, whereas in geography, critical reflection of the enquiry is mentioned only in the attainment target and then only from level 7 upwards (see Figure 3, page 12). The differences in emphases reflect the concerns of those who constructed the national curriculum documents rather than essential differences between subjects.

What makes enquiry distinctly geographical is related first to the subject matter to which the enquiry process is applied and second to the range of questions that have been developed by geographers to investigate the world (see Chapter 3).

> *Figure 6: Developing and applying the five thinking skills. Source: DfEE, 1999b, p. 23.*
>
> **Thinking skills**
>
> By using thinking skills pupils can focus on 'knowing how' as well as 'knowing what' – learning how to learn. The following thinking skills complement the key skills and are embedded in the national curriculum.
>
> **Information-processing skills**
>
> These enable pupils to locate and collect relevant information, to sort, classify, sequence, compare and contrast, and to analyse part/whole relationships.
>
> **Reasoning skills**
>
> These enable pupils to give reasons for opinions and actions, to draw inferences and make deductions, to use precise language to explain what they think, and to make judgements and decisions informed by reasons or evidence.
>
> **Enquiry skills**
>
> These enable pupils to ask relevant questions, to pose and define problems, to plan what to do and how to research, to predict outcomes and anticipate consequences, and to test conclusions and improve ideas.
>
> **Creative thinking skills**
>
> These enable pupils to generate and extend ideas, to suggest hypotheses, to apply imagination, and to look for alternative innovative outcomes.
>
> **Evaluation skills**
>
> These enable pupils to evaluate information, to judge the value of what they read, hear and do, to develop criteria for judging the value of their own and others' work or ideas, and to have confidence in their judgements.

SUMMARY

What is written in the geography national curriculum related to enquiry is relatively brief. The different ways we have encountered enquiry in our professional lives have led to different understandings of enquiry and we bring these understandings to our reading of the national curriculum. The Schools Council geography projects have been very influential in developing some people's understanding of enquiry but each project emphasised different aspects of enquiry, leading to slightly different understandings. The School Council geography projects tended to encourage the development of some enquiry skills and to neglect others and this continues to influence how enquiry is developed in the classroom.

Enquiry is not something to be defined once and for all on paper. It is something to be developed in the classroom in particular school and curriculum contexts. Exactly how it is developed depends on professional judgements. These judgements will be underpinned by professional values, by what we think is really important in education. The question 'what is geographical enquiry?' is closely related to the question 'why should we develop enquiry?' This is the focus of the next chapter.

Figure 7: Enquiry across the curriculum at key stage 3. Extracts from National Curriculum Programmes of Study.					
Geography	Citizenship	History	ICT	Mathematics	Science
Ask geographical questions and identify issues Suggest appropriate sequences of investigation	Think about topical political, spiritual, social and cultural studies		Be systematic in considering the information they need	Specify the problem and plan	Use scientific knowledge and understanding to turn ideas into a form that can be investigated and to decide on an appropriate approach
Collect, record and present evidence		Identify, select and use a range of appropriate sources of information	Obtain information well matched to the purpose by selecting appropriate sources	Collect data from a variety of suitable sources Process and represent the data	Make observations and measurements
Analyse and evaluate evidence Appreciate the influence of values and attitudes on issues	Analyse information and its sources, including ICT-based sources	Evaluate the sources used Select and record information relevant to the enquiry	Interpret information and represent it in a variety of forms that are fit for the purpose	Interpret and discuss the data	Use a wide range of methods to represent and communicate qualitative and quantitative data Describe patterns of relationships
Draw and justify conclusions Communicate in ways appropriate to the task and audience	Justify orally and in writing a personal opinion about such issues, problems or events	Reach conclusions		Answer the initial question by drawing conclusions from the data	Use observations, measurements and other data to draw conclusions
			Question the plausibility and the value of the information found	Examine critically and justify their choice of mathematical presentation of problems involving data	Consider whether the evidence is sufficient to support any conclusions or interpretations made

*All other things being equal,
one activity is more
worthwhile than another if it
asks students to engage in
enquiry into ideas, applications
of intellectual processes, or
current problems either
personal or social'*

(Raths, 1997, p. 67).

INTRODUCTION

At a recent conference someone commented to me that teachers were always being exhorted to use an enquiry approach but without any explanation of why such an approach should be used. Such exhortations now have 'official' support in the statutory requirements for geography and in the agenda for Ofsted inspections. Enquiry is presented as a good thing. But why?

This chapter sets out to explore justifications for the use of 'enquiry' in geographical education. Justification of educational practice should be related to fundamental beliefs and values about education rather than on what 'works' best in terms of test and examination results. We cannot determine what works best without first considering what we want to achieve; we need to justify what happens in the classroom in terms of what we believe in and then develop it so that it 'works'.

The first part of this chapter looks at the reasons for using an enquiry approach in terms of theories of learning and support for learning. The second part looks at reasons for using an enquiry approach in terms of beliefs about the purposes of education.

HOW CHILDREN LEARN: SOCIAL CONSTRUCTIVISM

The ideas in this book are underpinned by a widely accepted theory of learning, namely constructivism (see panel). The central idea of constructivism is that we can learn about the world only through actively making sense of it for ourselves; knowledge cannot be transmitted to us ready made. As we do not live in isolation from each other, our knowledge of the world is said to be 'socially constructed' in the contexts in which we live. This means that we make sense of the world through the ways we are influenced by, and contribute to, the cultural contexts within which we live.

Constructivism.

Constructivism is a widely accepted theory of learning developed from the work of Piaget, Vygotsky, Bruner and other psychologists. Its central idea is that we can learn about the world only through actively making sense of it for ourselves; knowledge cannot be transmitted or delivered to us ready made (Barnes and Todd, 1995). The following ideas are related to this central idea of constructivism:

1. *How we see and understand the world depends on our existing ways of thinking.* We make sense of the world not with empty minds but with assumptions about how things are and how things work. We also have expectations and attitudes. These all influence what we see and hear and the sense we make of that information. We can only construct meaning in relation to what we already know.

2. *Each individual sees and understands the world differently* as each person has developed knowledge about the world through different experiences and different social and cultural encounters.

3. *In constructing new knowledge we are not adding separate new 'bits' of knowledge to what we already have,* like extensions to a building. To make sense of the new information we have to incorporate it into and reconstruct what we already know.

4. *Our constructions of the world are not fixed but are being modified continuously* as we experience new things and encounter new ways of thinking.

'Social constructivists' emphasise the role of other people in helping us to make sense of the world. The knowledge we have is not constructed in isolation from other people, but through interactions with others, e.g. in families, with friends and in groups to which we belong. We make sense of the

Constructivism ... continued

world through participating in the world and through sharing, discussing and debating how we understand things with other people. We develop some common understandings, but because individual experiences are different, we all see the world slightly differently.

Some would argue that it is not enough to take account of interactions (Jackson, 1987). Within any groups, societies and cultures, some ways of seeing the world and of constructing knowledge are given preference over others. Ways of understanding which become dominant can influence the social structures within which we live. What we encounter in our lives is limited by the cultures in which we participate and by how the world is presented to us within our cultures through other people, institutions and the media. Dominant ways of seeing can restrict the development of alternative ways of seeing.

Although the theory of constructivism emphasises the world in our minds, the world we have constructed, this does not mean that any construction of the world is valid or possible. Although different individuals and groups see the world differently, this does not mean that anything goes. We do have to make sense of the world in relation to the material world as it can be known from our perceptions.

If we accept this theory of learning, then we also need to accept that students cannot learn simply by having ideas transmitted to them; they have to be actively engaged in the construction of geographical knowledge. Acceptance of the social constructivist theory of learning has implications for how enquiry work should be developed. We need to:

- take account of students' existing knowledge and ways of understanding;
- allow time for students to explore new information and to relate it to what they already know – making sense is not an instant process;
- provide opportunities for students to reshape and reconstruct their existing knowledge in light of new knowledge in discussion with others;
- make students aware of the way they see things and make them aware of different ways of seeing things;
- make students aware that all geographical knowledge has been constructed. What we know geographically has been constructed by people who have asked particular questions arising from interests at particular times and in particular places. Geographers have developed ways of seeing the world and constructing theories to understand it. Driver *et al.* argue that if science students were to be given access to the ways scientists understand the world, then 'they need to go beyond personal empirical enquiry' (1994, p. 6). Similarly, if students are to be given access to geographers' ways of understanding, then they need to be introduced to the concepts and models of the subject and learn to use these concepts and models themselves.

SUPPORT FOR LEARNING IN THE 'ZONE OF PROXIMAL DEVELOPMENT'

How are students to be helped to construct knowledge? The Russian psychologist Vygotsky's (1962) work has been influential on the development of the constructivist theory of learning, particularly on the need to support learning. Vygotsky became interested, through his experimental work with young children, in how they could solve problems beyond their existing levels of understanding if they were given 'slight assistance'. He referred to the difference between what a child could do without help and what he or she could do with support as the 'zone of proximal development' (see panel opposite). Vygotsky's assumption was that children would eventually be able to solve the problems unaided.

The zone of proximal development (ZPD).

In the 1920s and 1930s, Vygotsky carried out a series of studies to investigate how children developed their conceptual thinking. He was interested in the relationship between a child's level of development and the learning of concepts. He used problem-solving tests to find out children's mental ages, i.e. existing levels of development. In the research up to that stage, the tests were considered invalid if the children had been given any assistance. Vygotsky, however, questioned this and considered that finding out a child's mental age should be the starting point rather than the finishing point of study. In *Thought and Language* Vygotsky wrote,

'We tried a different approach. Having found that the mental age of two children was, let us say, eight, we gave each of them harder problems than he [sic] could manage on his own and provided some light assistance: the first step in a solution, a leading question, or some other form of help. We discovered that one child could, in co-operation, solve problems designed for 12 year olds, while the other could not go beyond problems intended for 9 year olds. The discrepancy between a child's actual mental age and the level he reaches in solving problems with assistance indicates the zone of his proximal development; in our example, this zone is four for the first child and one for the second. Can we truly say that their mental development is the same?' (Vygotsky, 1962, p. 103),

Several key aspects of the ZPD can be identified from the above quotation.

- The ZPD is ahead of what a child can already achieve unaided.

- The ZPD has limits; there is an area beyond it where problems are too hard for children to solve even with assistance.

- Each individual child has his or her own ZPD in which learning can take place with assistance. There can be big discrepancies in what children can achieve with support.

Vygotsky's ideas, developed in relation to the concept of the ZPD are significant for teaching and learning. (Quotations from Vygotsky, 1962, p. 104.)

- 'The only good kind of instruction is that which marches ahead of development and leads it.' If children are given problems to do that they can already handle without help then this fails to use the ZPD and they will fail to learn new things. Children are capable of learning new things within the ZPD. Vygotsky described them as being 'ripe' to learn.

- A child can be expected to progress but 'only within the limits set by the state of his development'. It is important that work is not beyond the ZPD, i.e. beyond what children can achieve with help.

- 'With assistance, every child can do more than he can by himself.' Children's conceptual learning is developed in the ZPD in collaboration with an adult or a more competent peer.

- 'What the child can do in co-operation today he can do alone tomorrow.' It is important that children should ultimately be able to do the work independently.

Vygotsky's (1962) concept of the zone of proximal development has implications for geographical enquiry.

1. Enquiry work should be beyond what students can already do, beyond their zone of previous development (Figure 1).

2. Enquiry work needs to help students to progress by providing them with challenges beyond what they can already do, i.e. within their zones of proximal development. These will be different for different students in a class.

3. Students need to be supported in enquiry work to enable them to achieve higher levels.

4. Enquiry work should not be beyond their zones of proximal development, beyond what they are capable of achieving at a particular time.

Vygotsky's ideas suggest that teachers have a crucial role to play in supporting enquiry learning. His ideas challenge the notion of totally independent enquiry work; they suggest that whenever students work with new conceptual frameworks beyond their present levels of thinking they will need teacher support to achieve higher levels of thinking.

Figure 1: The zone of proximal development illustrated by examples of two children.

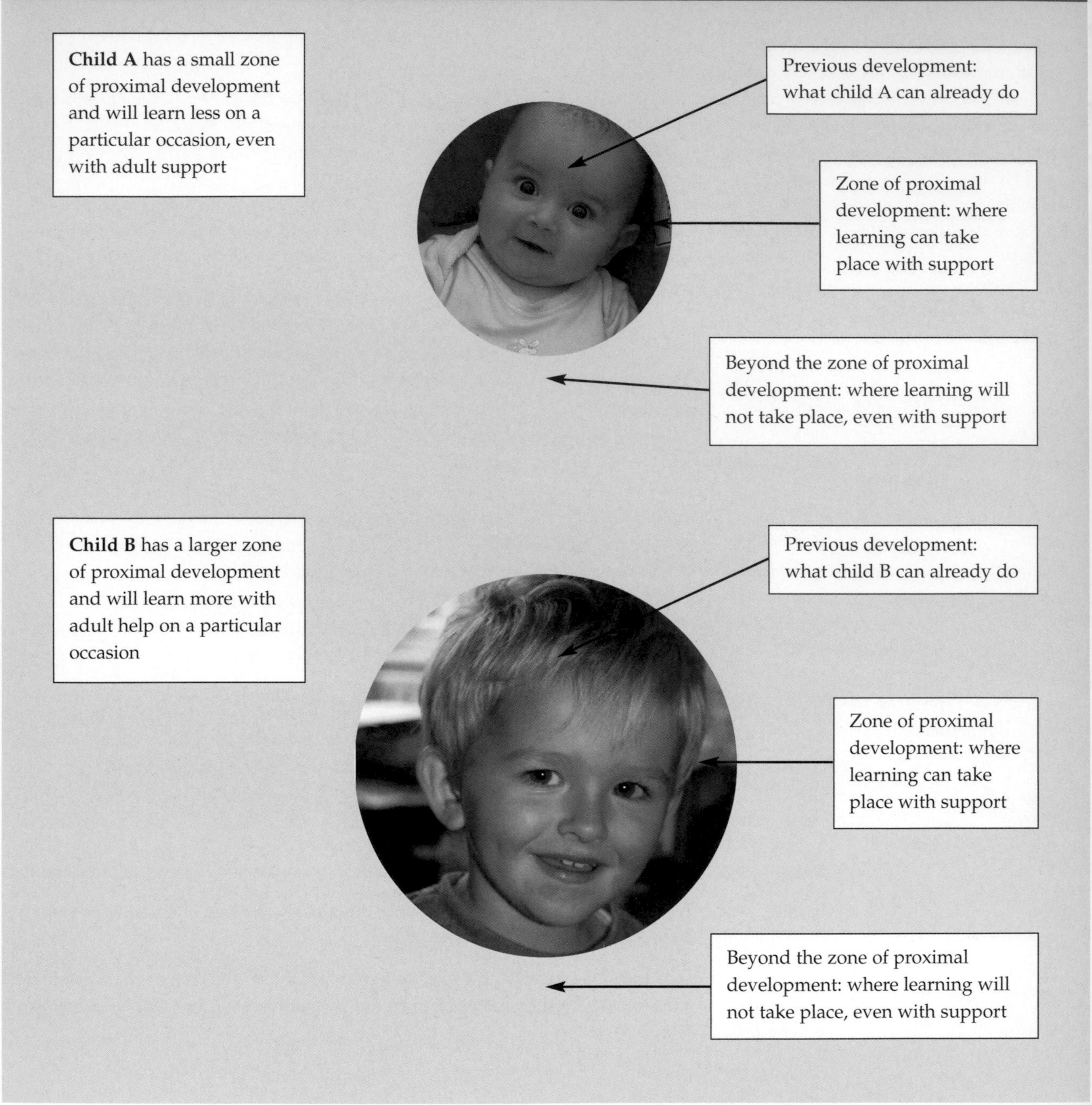

Child A has a small zone of proximal development and will learn less on a particular occasion, even with adult support

Previous development: what child A can already do

Zone of proximal development: where learning can take place with support

Beyond the zone of proximal development: where learning will not take place, even with support

Child B has a larger zone of proximal development and will learn more with adult help on a particular occasion

Previous development: what child B can already do

Zone of proximal development: where learning can take place with support

Beyond the zone of proximal development: where learning will not take place, even with support

SCAFFOLDING

Vygotsky's (1962) ideas about support for learning have been developed by many psychologists. What he termed 'light assistance' is now sometimes referred to as 'scaffolding' (see panel). In its current educational usage, 'scaffolding' refers to two different aspects of teacher support:

- support devised by the teacher at the planning stage
- support provided by the teacher during activities.

Scaffolding: 'the difference that makes the difference'.

Although Vygotsky did not use the term himself, he was the first to develop the concept of 'scaffolding'. In his experimental work with children, he studied how children's concepts could be developed in collaboration with an adult. He wrote about adults giving 'light assistance' to enable children to handle problems that they could not solve on their own. In *Thought and Language* Vygotsky (1962) referred to several types of assistance that might be given when a child was engaged in a problem-solving activity:

- providing the first step in a solution
- asking a leading question
- explaining
- supplying information
- questioning
- correcting
- making the child explain

Wood *et al.* (1976) were the first to use the word 'scaffolding' in an educational context. They carried out experimental work with young children to investigate 'the nature of the tutorial process: the means whereby an adult or "expert" helps somebody who is less adult or less expert' (Wood *et al.*, 1976, p. 90). They described 'the process that enables a child or novice to solve a problem, carry out a task or achieve a goal which would be beyond their unassisted efforts' as 'a kind of scaffolding'. They designed a task in which children had to build a pyramid out of building blocks, a task designed to be challenging, but not so difficult as to lie completely beyond the capability of any of the children and where the learning could potentially be applied to a later activity. In the experiments the tutors allowed the children to do as much as possible themselves and established an 'atmosphere of approval'. They collected data on the interaction of the tutor and the learners and found that scaffolding involved much more than modelling and imitation. They found that tutors 'scaffolded' the activities through dialogue and intervention by:

- reducing the number of steps involved in a task by simplifying the task;
- maintaining the pursuit of the goal and helping the learners to risk a next step;
- noting inconsistencies between what the child had produced and the ideal solution;

- controlling frustration and risk during the problem-solving activity, but without creating too much dependency on the tutor;
- demonstrating an idealised version.

Webster *et al.* (1996) investigated and developed ideas about scaffolding as part of a large-scale research project into the development of children's literacy. They collected detailed data of tasks and interactions in year 6 and year 7 classrooms (10-12 year olds). Although they recognised the importance of teachers devising tasks appropriate to students' needs, they found that a study of the tasks was insufficient to determine what or how children learnt. Different teachers mediated the same tasks in different ways. They found that 'the most powerful determinant of children's learning, the difference which makes the difference' (Webster *et al.*, 1996, p. 151) was how teachers 'scaffolded' the learning process. By 'scaffolding' they meant 'the complex set of interactions through which adults guide and promote children's thinking'. This definition emphasised that 'scaffolding' was much more than teachers simply providing help. It was a collaborative process involving dialogue in which learners had as important a role as teachers. Webster *et al.* concluded that scaffolding was 'the critical link between teacher and child' (1996, p. 96). From their research data, they identified various components of teaching and learning when these critical interactions and intervention took place:

- getting children involved in the task
- helping children to represent tasks in terms they understand
- helping children to adapt and develop concepts
- helping children externalise their learning. Listening to how children are pursuing the learning activities
- reviewing the process of learning and its worth

The research data showed that the nature of scaffolding varied from teacher to teacher. They concluded that: 'in order to be good at scaffolding teachers must have a precise knowledge of the characteristics and starting point of the learner, together with a thorough knowledge of the field of enquiry' (Webster *et al.*, 1996, p. 151).

Although the research studies referred to above were carried out at different times and in different contexts, they all emphasise certain aspects of scaffolding:

1. Scaffolding is a collaborative interactive process between teacher and learner involving dialogue.

2. Scaffolding is aimed at enabling learners to attain higher levels of understanding than they could unaided.

3. The ultimate aim of scaffolding is to enable learners to carry out the activity independently without the need for scaffolding.

At the planning stage teachers can scaffold the construction of new knowledge by:

*   simplifying complex subject matter
*   providing conceptual frameworks
*   planning an appropriate sequence of learning.

It is likely that some students will need more support than others. The ultimate aim would be for students to be able to use similar conceptual frameworks and plan sequences of investigation work independently. Students would, however, need support again to use new conceptual frameworks in more complex enquiry work.

Scaffolding can be provided through both whole-class and individual discussion. In order to support a student's construction of knowledge teachers need to get into the minds of the learner, they need to know and understand what the student is thinking and they need to be aware of misunderstandings. In order to be good at scaffolding, teachers need to have 'a precise knowledge of the characteristics and starting point of the learner, together with a thorough knowledge of the field of enquiry' (Webster *et al.*, 1996, p. 151). To be good at scaffolding teachers need to listen to students as much as talk to them. Teachers can begin to understand the student's perspective on their learning, for example, by asking them to:

*   explain what they are doing and why
*   explain a concept
*   review what they have just written.

There is potentially more opportunity for such interactive dialogues in lessons in which students are engaged in activities than in transmission-style lessons.

A BRIEF DIVERSION: METAPHORS, MAPS AND JOURNEYS

'Scaffolding' is a useful metaphor. Scaffolding is put up before and during the building process and is designed to support the construction of a particular building. Scaffolding is a temporary feature; it is removed when the building is complete. When a new building is built, more scaffolding is needed. Educational scaffolding supports students to achieve higher levels of understanding than they would without support. The ultimate goal is to enable students to achieve these levels of achievement independently.

If geographers had thought of a metaphor for support, they might well have come up with the idea of maps or journeys instead of scaffolding. If we need to go somewhere we have never been before, someone who knows the way could take us. We might not, however, learn how to get there on our own. To help us learn to make the journey independently we can be supported by a map. We can learn to use the map independently and eventually the map might become internalised in our minds; we have a mental map of how to make the journey. If, however, we want to make a new journey we need the support of a new map.

EDUCATIONAL AIMS

An educational practice can be justified in terms of whether the knowledge gained from it is worthwhile and in terms of whether it contributes to the achievement of more general educational purposes. What we consider as worthwhile knowledge and worthwhile purposes depends on what we value most. It is a matter of professional judgement.

Enquiry-based learning involves a change in teaching and learning styles and this changes not only how students learn but also what they learn (Roberts, 1995). To justify the change we must value what is learnt in the new style of teaching and learning more than what was learnt in the old style. The Schools Council geography projects justified their shift to 'enquiry-based learning' in terms of what they valued educationally. The projects wanted to replace a transmission style of teaching, in which the teacher was 'the source of information and ideas', to one in which the students learnt through activities. This was because they put greater value on students understanding and applying the fundamental concepts of the subject than on 'memorising' and 'factual recall'. They put value on students thinking for themselves, making their own judgements and forming their own opinions rather than accepting 'passively the conclusions, research and opinions of others'. They put value on students developing a range of general skills in addition to the development of specifically geographical skills, such as map interpretation. (See pages 17-22 for sources of quotations.)

In advocating enquiry-based learning, the Schools Council geography projects (see Chapter 1, pages 17-22) were also interested in how this could contribute to the achievement of broader educational purposes. The projects were not only concerned with academic progress for its own sake, they also wanted to contribute to the 'ordinary lives' of the students. It is interesting that the projects stated this not in vocational terms but in terms of contributing to students' capacity to make sense of the world they lived in.

The Schools Council projects also saw enquiry-based learning as a way of empowering students. They wanted to give them more control over their learning and they wanted students to gain the confidence and skills to become involved in discussing 'real world' issues, not only in the classroom but also as future citizens.

Figure 2: The participation dimension. Source: Barnes et al., 1987.

	◄——— Closed ———	——— Framed ———	——— Negotiated ——►
Content	Tightly controlled by teacher. Not negotiable.	Teacher controls topic, frames of reference and tasks; criteria made explicit.	Discussed at each point; joint decisions.
Focus	Authoritative knowledge and skills; simplified, monolithic.	Stress on empirical testing; processes chosen by teacher; some legitimation of student ideas.	Search for justifications and principles; strong legitimation of student ideas.
Students' role	Acceptance; routine performance; little access to principles.	Join in teacher's thinking; make hypotheses, set up tests; operate teacher's frame.	Discuss goals and methods critically; share responsibility for frame and criteria.
Key concepts	'Authority': the proper procedures and the right answers.	'Access': to skills, processes, criteria.	'Relevance': critical discussion of students' priorities.
Methods	Exposition: worksheets (closed); note giving; individual exercises; routine practical work. Teacher evaluates.	Exposition, with discussion eliciting suggestion; individual/group problem solving; lists of tasks given; discussion of outcomes, but teacher adjudicates.	Group and class discussion and decision making about goals and criteria. Students plan and carry out work, make presentations, evaluate success.

The way we develop geographical enquiry embodies our values. The meaning of enquiry in our own practice will depend on the professional judgements we make on what types of knowledge we value and what we consider is the purpose of education.

One significant way in which geographical enquiry work can vary is the extent to which teachers or students control the learning. It is worth examining this in more detail.

WHOSE ENQUIRY?

Barnes *et al.* (1987), in their evaluation of the national Technical and Vocational Educational Initiative (TVEI) Project, devised a framework to analyse the extent of student participation in TVEI lessons (Figure 2). The categories at the top of the framework indicate the extent of student participation. Although Barnes *et al.* considered that the nature of participation could be indicated along a continuous dimension, they divided the dimension into three categories: closed, framed, and negotiated.

This framework has been adapted to make it applicable to geographical enquiry work (Figure 3). In Figure 3, at the closed style end of the framework everything is controlled by the teacher: the questions to be asked; the resources to be used; the skills to be used; the conclusions to be reached; and how the enquiry is to be evaluated. Furthermore, the reasons for the choices that have been made are not made explicit and are not discussed with students. Whether this style of teaching is considered to represent a teacher-controlled enquiry or whether it cannot be called enquiry at all is a matter of debate. It depends on what one considers are the essential elements of enquiry. The students would have been passengers on some sort of enquiry journey. They might have developed some enquiry ski

	← ——— Closed ———	——— Framed ———	——— Negotiated ——→
Content	Focus of enquiry chosen by teacher.	Focus of enquiry chosen by students within theme (e.g. choosing which volcano to study).	Student chooses focus of enquiry (e.g. choosing which less economically developed country to investigate).
Questions	Enquiry questions and sub-questions chosen by teacher.	Teacher devises activities to encourage students to identify questions or sub-questions.	Students devise questions and plan how to investigate them.
Data	All data chosen by teacher. Data presented as authoritative evidence.	Teacher provides variety of resources from which students select data using explicit criteria. Students encouraged to question data.	Students search for sources of data and select relevant data from sources in and out of school. Students encouraged to be critical of data.
Making sense of data	Activities devised by teacher to achieve pre-determined objectives. Students follow instructions.	Students introduced to different techniques and conceptual frameworks and learn to use them selectively. Students may reach different conclusions.	Students choose their own methods of interpretation and analysis. Students reach their own conclusions and make their own judgements about the issue.
Summary	The teacher controls the construction of knowledge by making all decisions about data, activities and conclusions.	The teacher inducts students into the ways in which geographical knowledge is constructed. Students are made aware of choices and are encouraged to be critical.	Students are enabled, with teacher guidance, to investigate questions of interest to themselves and to be able to evaluate their investigation critically.

Figure 3: The participation dimension in geographical enquiry.

There is much greater student participation in the framed category of enquiry work. Although the teacher controls the overall framework, perhaps determining the key questions, selecting the data and possible ways of interpreting it, the students are let into the secret of what it means to construct geographical knowledge. They are aware of the choices being made by being brought into the discussion. There is scope for reaching their own conclusions and of thinking for themselves. This style of working, which could be termed 'framed enquiry', enables teachers to meet the requirement of the GNC in terms of content and assessment demands and at the same time it develops many aspects of enquiry learning.

In the negotiated category of enquiry work, control is largely handed over to the student. There is still support from the teacher, but it is the student who makes the key enquiry decisions. It is the student who determines the focus of the enquiry, the questions to be asked, the selection of data, the methods of analysis and interpretation and what conclusions are reached. In this style, the student has opportunity to use the complete range of enquiry skills.

These categories oversimplify the many ways in which teaching and learning styles can differ from lesson to lesson even with the same teacher and the same class. Some classroom practice will fall somewhere between two of these styles. More often all three styles are used at different stages of an investigation. For example, a teacher might start enquiry work in a closed way, choosing the focus of enquiry, the questions and the resources, but enable students to make choices about how to present and analyse data (Roberts, 1996). Alternatively, teachers might allow students to choose the focus of their enquiry (e.g. which volcano or country to study) but provide structured guidelines on how this should be investigated.

The value of the framework is in making us more aware of the professional judgements we are making in planning enquiry work, aware of which aspects of learning we control, how we share our decisions with students and the extent to which we are willing or it is appropriate to negotiate. The professional decisions are not only about geography, they are about enabling students to become critically engaged in constructing their own world views.

SUMMARY

An enquiry approach to learning is consistent with a widely held theory of learning. Enquiry can be justified because of the emphasis it places on thinking and understanding, rather than on memorisation. Enquiry can also be justified because it can be used to achieve broader educational purposes. The particular way in which geography departments and individual teachers develop enquiry work will embody what is valued. If we think it is important that enquiry work should be relevant to students' present and future lives, then the issues studied should be relevant and important to students themselves. If we think it is important that enquiry work should empower students to take more control of their own learning and their own lives, then we need to be aware of when and how and why we are controlling enquiry work and when and why and how we are enabling students to have control. The planning of enquiry work involves many professional judgements and these will determine the meaning of geographical enquiry in particular contexts.

'To hear people talking about facts, you would think that they lie about like pieces of gold ore in the Yukon days waiting to be picked up. There are no such facts. Or if there are, they are meaningless and entirely ineffective; they might, in fact, just as well not be lying about at all until the prospector — the journalist — puts them into relation with other facts: presents them in other words. They become as much a part of a pattern created by him [sic] as if he were writing a novel. In that sense all stories are written backwards. They are supposed to begin with the facts and develop from there, but in reality they begin with a journalist's point of view, a conception'

(Cockburn, quoted in Wheen, 2002, p. xii).

INTRODUCTION

Research evidence suggests that enquiry-based learning is not well developed in the classroom (QCA, 1998; Roberts, 1998; Davidson and Catling, 2000). The aim of this chapter is to encourage a fresh look at geographical enquiry to support its further development at key stage 3. It starts by considering the kinds of questions geographers ask. Next, it identifies and discusses four essential aspects of geographical enquiry: a need to know; using data as evidence; making sense of data; and reflecting on learning. It then presents a framework for learning through enquiry in which the essential aspects are related to a wide range of thinking skills. Finally, the chapter introduces the categories of enquiry used to structure the second part of this book.

WHAT ARE THE CORE QUESTIONS OF GEOGRAPHY?

The GNC requires that students are taught to ask 'geographical questions'. What makes a question 'geographical'? The questions that geographers ask are one of the distinguishing features of the discipline. They frame the way geographers look at the world: they influence how geographical knowledge is constructed. Whether teachers devise the overarching questions for enquiry work at key stage 3 or whether they encourage students to develop their own questions, it is necessary to consider what constitutes a 'geographical' question.

As geography has changed as a discipline, so have the core questions changed. Neighbour, in discussing an overview of the development of geographical thought, identified the emergence of five core questions:

1. 'What is the phenomenon?

2. Where is it located?

3. Why is it located there?

4. What impact does its location have?

5. What changes should be made? What ought to be done?' (1992, p. 15).

The questions *what* and *where* have always dominated descriptive approaches to geography. During the second half of the twentieth century, geographers became increasingly interested in answering the question *why there*, at first in terms of formulating scientific laws and later in terms of people's preferences and perceptions. Until the 1970s, the core questions of geography were *what* and *where* and *why* and *how*. It was not until the development of a welfare approach to geography (Smith, 1977), with its emphasis on social relevance, that new questions emerged as core questions: *with what impact* and *what ought*.

Neighbour claimed that 'the questions *what, where, why, with what impact* and *what ought* have received national and international recognition as the focus for geographical education at high school level' (Neighbour, 1992, p. 15). Although Neighbour's five core questions are not listed as such in the GNC (DfEE, 1999a) the *what, where* and *why* questions are implicit in what is set out in the programmes of study and the attainment target. The GNC draws particular attention to the importance of *what impact* and *what ought* questions by listing them as exemplars:

'Pupils should be taught to: ask geographical questions (for example, 'How and why is this landscape changing?; What is the impact of the changes?; What do I think about them?)' (DfEE, 1999a, p. 22).

The questions that are implicit and explicit in the GNC can be traced back to the Schools Council's 'route for enquiry' sequence of questions and these can provide a useful framework for planning enquiry at key stage 3:

- What?
- What and where?
- How? Why?
- What might? What will? With what impact?
- What decision? With what impact?
- What do I think? Why?
- What next? What shall I do?

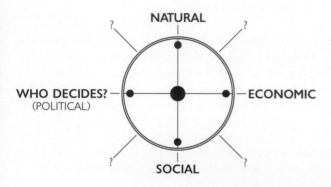

The Compass Rose (Birmingham DEC, 1995) provides a different way of categorising geographical questions. Instead of setting out different types of questions, it encourages questions on different aspects of geography:

- **Natural:** questions about the natural environment.
- **Economic:** questions about money, trade and aid.
- **Social:** questions about people and their relationships.
- **Who decides:** questions about power and who decides and about who benefits and who loses from political decisions.

Although the Compass Rose questions are different from those in Neighbour's list of core questions, they can both be used to support geographical enquiry at key stage 3. Different sets of questions encourage different types of thinking. A framework of questions can support enquiry, but it can also limit it because it focuses attention on some aspects of a theme or place and neglects others. Questions can help us develop different ways of seeing, but also different ways of being blinkered.

One way in which the questions above limit thinking is that they do not incorporate some of the changes taking place in the subject in higher education. Recent developments in human geography emphasise the subjective nature of geography and the selective, provisional nature of geographical knowledge (Jackson, 2000). There has been a growing interest in the way different people experience the world, how their place in the world is represented by others and the ways in which their worlds are inter-related. There is scope for the curriculum at key stage 3 to incorporate these perspectives, with enquiry work framed by an additional core question:

- How do different people experience space and place?

Aspects of what this question is seeking to explore are included in some of the examples in the second half of this book, particularly in Chapters 13 and 14.

Whereas curriculum development in geography has tended to focus on general enquiry questions and their application to varied themes and places, historians have devised 'big' specific questions to frame the study of a unique event. Riley (1999) argues that the 'carefully crafted enquiry question' is an important 'pedagogic tool' in the teaching of history. The questions he devised for schemes of work for history at key stage 3 are more specific to the context than many of the questions asked in geography, for example:

- Is Farleigh Hungerford a typical medieval castle? (Y7)
- Does Oliver Cromwell deserve his reputation as a harsh dictator? (Y7)

- Did life get better for ordinary people? (Y8)
- Why did it take so long for women to get the vote? (Y8)
- Why do historians disagree about the causes of the Second World War? (Y9)
- How should we remember the Holocaust? (Y9)

The contrast between these questions and the core questions of geography is challenging. Do the kinds of questions asked in history provoke more interest among students? If the same general questions are used repeatedly in geography do these become tedious? Is it indeed appropriate to use the same general questions to investigate themes and places that are very different? Can South Africa, with its unique history and development and range of issues, be investigated through the same range of questions as those applied to any other country? Perhaps the general questions identified by Neighbour (1992), the 16-19 Geography Project, and the Compass Rose are not enough. Perhaps there is need to devise specific geographical questions that would satisfy Riley's criteria of 'good' enquiry questions. He asked history teachers to consider what makes an enquiry question a good one by asking themselves these questions:

Does each of your enquiry questions:

- *Capture the interest and imagination of your pupils?*
- *Place an aspect of historical thinking, concept or process at the forefront of the pupils' minds?*
- *Result in a tangible, lively, substantial, enjoyable "outcome activity through which pupils can genuinely answer the enquiry question"' (Riley, 1999, p. 8).*

A NEED TO KNOW

An enquiry approach to learning recognises that knowledge is not something 'out there' ready to be learnt; it is generated in the process of answering questions. All geographical knowledge has been generated by someone who has, at some time, in some place, been puzzled and has wanted to know and understand more. Geographers have asked questions because they have been stimulated by a sense of curiosity, by their speculations on how and why things are as they are and by their having a 'need to know'.

Questions have always been seen as an important part of geographical enquiry. Naish *et al.* stated, in relation to the Schools Council 16-19 Geography Project, that the starting point for enquiry was 'necessarily a question, problem or issue' (1987, p. 36). Davidson and Catling (2000), although emphasising the centrality of questioning to enquiry, argued that questions were not the starting point; they need to be preceded by some kind of stimulus. They suggested a useful list of 'stimulus ideas' including the use of photographs, video, music, newspaper articles, fiction, brochures, food tasting, advertisements, games and cartoons to provoke questions from students. They wrote, 'the key principle for using stimulus materials is to encourage pupils to ask questions and identify issues that are worth studying' (Davidson and Catling, 2000, pp. 281-2). The level descriptions in the GNC also acknowledge that questions are not the starting point but that they arise out of knowledge and understanding: 'Drawing on their knowledge and understanding, [students] suggest suitable geographical questions' (DfEE, 1999a, Level 4).

Questions are frequently used to structure what is studied in geography. Questions have been used, for example, to structure:

- specifications for some GCSE and A-level geography examinations
- *Geography: A scheme of work for key stage 3* (QCA/DfEE, 2000)
- chapter headings in some geography textbooks
- some schemes of work produced by geography departments.

The use of questions in planning frameworks is generally presented as good practice, but this does not necessarily ensure that what happens in the classroom is enquiry based (Davidson and Catling, 2000). The questions can become content headings rather than questions which really perplex and create a need to know. Second, these questions are pre-determined for students; they do not necessarily generate the students' own questions.

Questions need to become more than routine content headings. The questions need to become the students' own questions, even where the 'big' questions have been pre-determined. Students need to develop what I think of as 'question hooks' in their minds, formed from their curiosity and a need to know. This can be a challenging task. Many of the themes in the GNC are distant from the everyday lives and experiences of students. Most key stage 3 students do not come into the geography classroom eager to know about what they are required 'to be taught' by the GNC programmes of study; they do not necessarily want to know about, for example, flood hydrographs, Kenya, or life expectancy rates.

There are several ways of addressing this challenge:

1. Teachers can provide the kinds of stimulus suggested above to provoke questions. Starter activities need to be connected in some way to what is going to be studied and need to encourage speculation, hypothesising, and the asking of questions.

Elizabeth (then aged ten): 'I hate school, I can't stand teachers telling me things I don't want to know'

2. The teacher needs to convey a genuine interest in what is being studied, replicating in some way the curiosity that generated the knowledge in the first place. We need to recreate in the classroom the sense of the uncertainty that preceded what is now known and the spirit of enquiry that provoked the questions. Bruner (1986) quotes a teacher who said in class, 'It is a very puzzling thing not that water turns to ice at 32 degrees Fahrenheit, but that it should change from a liquid into a solid'. How often do teachers seem puzzled? I can remember my own biology teacher, Miss Page, wondering about how birds managed to migrate. She talked about it as a great mystery, as something she wanted to understand. How often do teachers wonder?

3. Enquiry work can be made relevant to the present and future lives of students by giving them the opportunity to link what they are studying to their own lives and ways of thinking about the world. They can be encouraged to express their own opinions.

4. Enquiry work can be organised so that students have some elements of choice in what they study and more control over their learning.

5. Where questions are pre-selected for enquiry work, as part of a planning framework, the questions presented to students should be 'good' questions. Good questions are those which really probe an issue or an area of study, are specific to what is being studied, and which are capable of generating many sub-questions from students.

'Needing to know' is not just an aspect of the first stage of enquiry work; it is important that the need to know is sustained throughout. Geographical enquiry is about having an inquisitive attitude towards the world and towards what we know and understand. It seems vital, if students are to learn anything in school, that they too have a 'need to know'. It is important that they too are made curious, are puzzled, and that they want to ask their own questions and develop lots of 'question hooks' throughout the investigation of a topic or issue.

Figure 1: *Types of data that can be used as evidence in classroom enquiry work at key stage 3.*

Text as data
Textbook descriptions and accounts
Non-fiction, e.g. travel writing
Fiction
Newspapers – articles, comment, letters
Magazines
Brochures
Advertisements

Visual data
Photographs
Paintings
Drawings
Diagrams
Advertisements
Video
Cartoons
Satellite images

Statistical data
Tables of figures
Bar graphs
Pie charts
Choropleth maps
Line graphs
Flood hydrographs

Population pyramids
Climate graphs

Maps
Ordnance Survey maps
Atlas maps – political, physical and thematic
Maps in use outside the classroom, e.g. in brochures and newspapers, football tickets
Weather maps

Personal knowledge
Memories of place including remembered images
Memories of events
Mental maps
Personal theories and ways of seeing
Personal knowledge gained indirectly

Objects
Bags of rubbish
Artefacts
Food
Rocks

USING DATA AS EVIDENCE

Another essential aspect of geographical enquiry work is the use of data as evidence rather than as fact. In the classroom only a small proportion of the data used can be collected first-hand by students through fieldwork or surveys. This is partly due to practical reasons and partly because not all themes lend themselves to first-hand data collection by students. So, necessarily, most data used for enquiry work in the classroom are secondary. This should not be seen as a limitation on classroom enquiry work as there is such a wide variety of data that can be used (Figure 1).

It is likely that for much enquiry work, even if students make some choices of their own, teachers would make a preliminary selection of data or sources of data. Selection could be informed by the following criteria:

- data should be accessible to the students in terms of literacy and numeracy
- data should be interesting
- data should be in a relatively unprocessed state, so that there is scope for students to make sense of the data for themselves
- data should include the kinds of geographical information students are likely to encounter outside the classroom, e.g. publicity maps, newspaper reports, and videos of television weather forecasts.

All geography textbooks contain some data satisfying these criteria, e.g. maps, photographs, statistics, and extracts from reports, newspapers or brochures. What is not so useful for enquiry work, either in textbooks or on teacher-produced resource sheets, are lists of bullet points or key ideas summarising the outcomes of somebody else's thinking. This kind of information provides students with what might well be desirable end-points of learning but without giving the students the opportunity to reach these conclusions through their own thinking.

Another almost inevitable disadvantage of data in textbooks is that only relevant data are provided; clearly, publishers are not going to print a lot of irrelevant data. If, however, students are to gain experience of locating and selecting data then they need to be able to distinguish between relevant and irrelevant data. The supported use of libraries and the internet provides opportunities for students to do this. In addition, resources provided by teachers can include data of various degrees of relevance. For example, mysteries usually include some statements that are 'deliberate red herrings … to encourage students to think hard about what information to use' (Leat and Nichols, 1999, p. 14). An example of an activity in which surplus data are provided is described in the Picture Editor strategy (see Example 1, Chapter 5, page 56) in which students, like photograph editors of newspapers, have to select relevant images from many. In this example, students were provided with 40 photographs related to the Mozambique floods from which they had to select only eight. We need to give students access to

Figure 2: Two children's very different ways of seeing patterns.

relevant, less relevant and irrelevant information if they are to develop the full range of enquiry skills. Only then can they become aware of the selective nature of the information we use to create geographical knowledge.

The selection of information from data is inevitably guided by the conceptual frameworks we use to determine what is and what is not selected. For example, if students watch a video without any teacher guidance, each student will remember different things. Students have their own individual ways of seeing the video, influenced by their knowledge, experiences and interests; their attention is drawn to different aspects of what they see. If we want to encourage geographical thinking, then we need to introduce students to geographical frameworks or ways of seeing. These frameworks will vary according to the purpose of the investigation. For example, students could select data from a video using categories such as 'social, economic, and environmental factors'. Alternatively, using the same video as a source of information, they could focus on gender differences. The data collected would be totally different. The influence of different ways of seeing can be illustrated, by way of analogy, by two completed patterns constructed by my daughters when they were very young (Figure 2). The pages of the design pad they were using were identical. Yet they picked out and shaded different patterns with strikingly different results. The point is that when we select and categorise data, even before we start to analyse it and interpret it, we are influenced by our ways of seeing.

This point was well made by Claud Cockburn in the quotation at the top of this chapter: journalists have a 'point of view', a 'conception', and this precedes the collection of 'facts' (Wheen, 2002). In the same way as journalists have points of view, so do geographers. When students are learning to study geography through enquiry, they are also learning geographers' 'points of view'. When students are selecting data, they are selecting according to some kind of geographical conception.

MAKING SENSE

Making sense is at the heart of learning. Geographical enquiry is essentially about using the information collected from data to develop understanding and to construct geographical knowledge. There is a difference between information and knowledge. Students can find information from a wide range of data but that in itself is not enquiry. Enquiry is not simply about finding information to answer questions; it is about developing understanding. In order to do this, students need to do something with the information they have gathered. They need to examine the data, to relate it to what they know already, to see relationships between different bits of information, to make all kinds of connections and to develop their own understanding of what they are studying.

The difference between using data and making sense of it, between finding answers and constructing knowledge, can be illustrated with an example.

Students could answer the question 'Where in the world do earthquakes occur?' by using a textbook which provides a list of generalisations about the distribution of earthquakes. They can transfer this information to their exercise book without really thinking much about the question or the answer. If they can copy accurately, they will have the 'correct' answer to the question. Alternatively, they can search for the data and try to make sense of it for themselves, e.g. using atlases or the internet (Example 4, Chapter 9, page 118). This way of answering the question *where* involves the students in constructing knowledge for themselves, rather than simply accepting knowledge constructed by others.

Geographical enquiry is about making sense. This aspect of enquiry is not limited to a later stage after the collection of data. Students need to make sense of what they are doing throughout an investigation. The critical role of literacy in enabling students to do this is examined in Chapters 4, 5, 6 and 7.

REFLECTING ON LEARNING

Another essential aspect of learning through enquiry is the process of reflecting on learning. This has two elements:

- reflecting on what has been learnt
- reflecting on how it has been learnt.

Reflecting on *what* has been learnt

Throughout enquiry work, it is important to stand back and ask critical questions. These might be in relation to the questions posed, the conceptual frameworks used to collect data, the validity and reliability of data sources and the data, the methods used to present and analyse data and the interpretations of the data. These should be open to scrutiny and evaluation. Students can be prompted to think critically about what they are doing throughout the enquiry process.

Following enquiry work, students need to look back on it and reflect on what they have achieved and whether they would have done things differently if they were carrying out the enquiry again. Reflecting on what has been learnt pays critical attention to the way knowledge has been constructed and could be part of all plenary debriefing activities.

Reflecting on *how* learning has taken place

Psychologists are paying increasing importance to the role of metacognition in learning. Bruner describes this as 'going meta' or 'turning around on what one has learned' (1996, p. 88). 'Going meta' involves making students aware of their own thought processes. When students are 'going meta' they are attempting to put their thinking processes into words, so that thinking processes can be shared, evaluated and developed further. Research has shown that students are capable of doing this and that 'going meta' improves learning (Bruner, 1996; Leat and Nichols, 1999). Encouraging students to 'go meta' can take place throughout enquiry work and need not be confined to summary debriefing activities. Questions to provoke reflection are included in many of the activities discussed later in this book.

Figure 3: A framework for learning through enquiry.

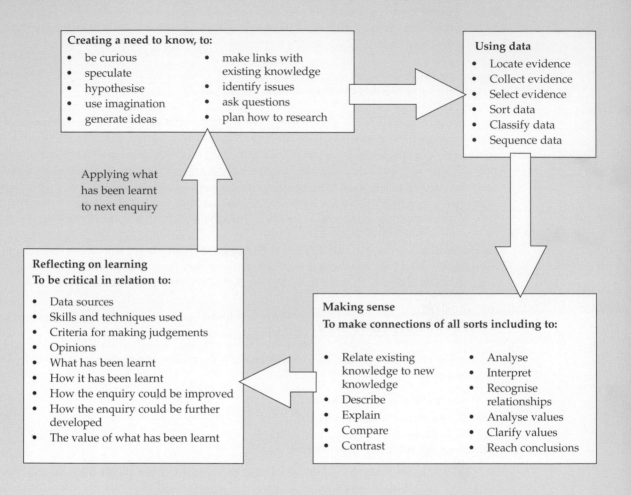

A FRAMEWORK FOR LEARNING THROUGH GEOGRAPHICAL ENQUIRY

The 'framework for learning through enquiry' presented and used throughout this book is based on the four essential aspects of enquiry, identified and discussed above, and on the kinds of thinking that might take place in relation to each aspect (Figure 3). The list of thinking processes has been derived from three sources which are identified separately in Figure 4:

1. the statutory requirements included in the programme of study for key stage 3 and attainment target for geography

2. all the activities included under the heading 'thinking skills' in the national curriculum handbook for secondary teachers (DfEE, 1999b)

3. activities associated with a constructivist theory of learning.

Although the thinking skills are set out in lists, this is not because they should be isolated and taught separately or as ends in themselves. The purpose of developing the skills is to increase students' capacity to learn, to know and to understand the world they live in and also their capacity to be critical about such knowledge.

Figure 4: Types of thinking involved in learning through enquiry.			
Aspect of enquiry	a. Activities from GNC key stage 3	b. Thinking skills from national curriculum handbook	c. Activities related to constructivism
Needing to know	• Ask questions • Identify issues • Suggest how to investigate questions • Suggest appropriate sequences of investigation	• Pose and define problems • Ask relevant questions • Generate ideas • Use imagination • Plan what to do and how to research • Suggest hypotheses • Predict outcomes • Anticipate consequences • Extend ideas	• Puzzle • Wonder • Speculate • Hypothesise • Make links with existing knowledge and experience
Using data	• Collect evidence • Select information and sources of evidence • Record data • Present evidence • Select skills and techniques	• Locate and collect relevant information • Sort • Classify • Sequence	• Relate data to existing conceptual framework • Be aware of new conceptual framework
Making sense of data	• Describe • Explain • Analyse evidence • Recognise relationships • Appreciate values and attitudes • Clarify and develop values and attitudes • Draw and justify conclusions • Communicate • Draw on geographical ideas and theories	• Compare • Contrast • Analyse part/whole relationships • Draw inferences • Make deductions • Make judgements and decisions informed by reasons or evidence • Make decisions • Test conclusions	• Re-order existing knowledge • Explore through discussion • Apply • Present
Reflecting on learning	• Evaluate critically sources of evidence • Evaluate • Evaluate critically • Suggest improvements • Suggest further lines of enquiry	• Evaluate information • Give reasons for opinions • Explain thinking • Use precise language to explain what they think • Judge value of what they read • Develop criteria for judging the value of their own and others' work • Look for alternative innovative outcomes • Improve ideas • Have confidence in judgements	• Reflect on what has been learnt • Reflect on how learning has taken place • Reflect on what is still puzzling • Form opinions

Figure 5: Categories of geographical enquiry.

- Describing (Chapter 9)
- Explaining (Chapter 10)
- Values (Chapter 11)
- Survey (Chapter 12)
- Personal geographies (Chapter 13)
- Representation (Chapter 14)
- Futures thinking (Chapter 15)
- The big project (Chapter 16)

A TYPOLOGY OF GEOGRAPHICAL ENQUIRY

The second half of this book is divided into eight chapters, each focusing on a different type of enquiry (Figure 5). There are several reasons for separating out what are normally considered to be integral parts of the same investigation.

First, my research into teachers' understandings of enquiry ('Research into practice', pages 14-16) found that some teachers thought that 'enquiry' was far too complicated for key stage 3. One teacher conceptualised enquiry almost entirely in relation to the 16-19 route for enquiry and considered that 'real enquiry' was not feasible at key stage 3. If teachers want to model their use of enquiry on the 16-19 route, then there are advantages to breaking it down into a series of separate enquiries. These can, of course, be combined in various ways to produce a unit of work so that the complete range of essential enquiry questions (see page 38) is addressed.

Second, by narrowing the scope of enquiry work, it is possible to carry out a complete enquiry from initial stimulus to the reaching of conclusions and debriefing within one or two lessons. Many schools have only one or two lessons of geography a week at key stage 3 and there are advantages in investigations which do not lose momentum over time.

Last, the categories of enquiry suggested in Figure 5 encourage a focus on different types of data, skills and strategies needed to answer different types of questions. For example, the skills needed to answer the question 'where?' are very different from those to answer questions such as 'what do I think about this issue and why?'.

Chapters 9, 10, 11 and 12 address most of the questions traditionally associated with enquiry frameworks at key stage 3. Chapters 13, 14 and 15 are included in recognition of changes in academic geography which seem relevant and worth developing at key stage 3. The rationale for each of these categories is discussed in the relevant chapters.

The categories are not presented as a definitive classification of types of enquiry. They are not intended to limit thinking by suggesting inflexible new boundaries. This would go against current thinking. Geographical knowledge is being advanced most vigorously in higher education on the boundaries of the subject, where the subject overlaps with other subjects and where old boundaries no longer limit thinking (Jackson, 2000). Indeed geographical enquiry can best flourish where questioning is not constrained and where all kinds of data can be used in developing new knowledge. However, at key stage 3, there could be advantages in focusing on particular types of questions and the data and skills needed to explore them. It is hoped that the categories will stimulate further thinking about enquiry, and encourage the kinds of curriculum development in which the different types of enquiry are combined in creative ways to form exciting units of work in which students have the opportunity to explore a wide range of geographical questions.

SUMMARY

This chapter presents a view of enquiry with four essential aspects. The aspects can be related to national curriculum requirements, thinking skills and to constructivism to produce a framework for learning through enquiry. This framework incorporates the use of a wide variety of separately identified enquiry and thinking skills. A typology of categories of enquiry has been presented, for development in the second part of this book.

INTRODUCTION

One of the elements of the National Strategy for Key Stage 3 is the promotion of 'Literacy across the Curriculum' (DfEE, 2001a). This part of the strategy is based on the idea that all teachers, whatever their subject, should contribute to the development of students' literacy and that all schools should develop policies to implement the proposals advocated by the strategy.

What is in this for geographical education? How does attention to literacy enhance geographical enquiry and understanding? These questions are the concerns of this and the next three chapters. This chapter outlines the contextual background to a growing interest in language and literacy across the curriculum and discusses three key ideas about language and literacy that seem to be of importance in developing geographical enquiry. These key ideas are then applied to one aspect of literacy: the development of vocabulary.

'Language enables us to share thoughts about new experiences and organize life together in ways in which no other species can'

(Mercer, 2000, p. 4).

THE BACKGROUND TO 'LANGUAGE ACROSS THE CURRICULUM'

Interest in language across the curriculum is not new. The present recommendations of the National Literacy Strategy seem, in some ways, remarkably similar to those of *A Language for Life*, often referred to as the Bullock Report (DES, 1975). Its fourth recommendation reads:

'Each school should have an organised policy for language across the curriculum, establishing every teacher's involvement in language and reading development throughout the years of schooling' (DES, 1975, p. 514).

The publication of the Bullock Report led to an increased interest in language across the curriculum by teachers and researchers. It had some impact in schools, but this depended on particular initiatives in schools and local education authorities. It has had a much bigger impact on research. Whereas previously, research into language and learning had been related mostly to the teaching of English, a substantial amount of research during the last 27 years, including major national projects, has focused on aspects of language across the curriculum.

Whereas the Bullock Report had its origins in concerns about reading standards, the current National Literacy Strategy had its origins in concerns mainly about standards of writing and particularly about the lower achievement of boys in standard assessment tasks (SATs). The purpose of the National Literacy Strategy is to 'raise standards', to be measured by achievement in SATs. There is emphasis, therefore, on what is assessed in SATs and the criteria by which it is assessed. The National Literacy Strategy gives more emphasis to reading and writing, both of which are assessed by SATs, than to speaking and listening which are not. The Bullock Report, in contrast, gave more attention to reading, but also gave attention to writing and the role of classroom talk (DES, 1975).

THREE KEY IDEAS

The chapters on literacy (4-8) relate to the National Literacy Strategy, but also draw on the wide range of research into language and learning that has taken place since the publication of the Bullock Report. Three key ideas from the research seem particularly relevant to the development of geographical enquiry:

1. Language is a means of learning.

2. In learning geography we are also learning the language of geography.

3. Teachers can create a classroom 'ecology' that promotes learning through language.

Each of these ideas is discussed below. These three key ideas inform the activities related to reading, writing and speaking and listening in Chapters 5, 6 and 7 and the activities related to different types of enquiry work in the second half of this book.

Key idea 1: Language is a means of learning

We use language not only to communicate with others but also to make sense of things for ourselves. Vygotsky, the Russian psychologist, emphasised the role of language in developing thought. In his research, he noted how talking out loud helped young children to solve problems. He believed that these external monologues were closely linked with the development of thinking and that they became internalised as children developed. He believed that thought 'comes into existence' through words:

'The relation of thought to word is not a thing, but a process, a continual movement back and forth from thought to word and from word to thought … Thought is not merely expressed in words: it comes into existence through them. Every thought tends to connect something with something else, to establish a relationship between things. Every thought moves, grows and develops, fulfils a function, solves a problem' (Vygotsky, 1962, p. 125).

While Vygotsky's words might seem quite abstract, we can probably all draw on personal experience in which we have developed our thoughts through discussion or through writing. When we contribute to discussion, we develop and shape our ideas, feelings and thoughts through attempting to put them into words. Even if we stay silent in discussion, we can still be exploring what we might say in our minds. This inner dialogue helps us to sort out what we think. Similarly, when we write, the process of writing in itself helps to clarify and develop ideas, feelings and thoughts. Writing can give shape to half-formed ideas. Vygotsky's (1962) work on language and learning has influenced much subsequent research into language and learning.

If we want students to learn through geographical enquiry, then we need to provide them with opportunities to use language to shape their own thinking as well as to communicate with others through talk and writing. We must allow time for students to develop their thinking through use of tentative, exploratory forms of talk and writing. This means devising strategies that value the process of sorting out ideas and provide opportunities for students to 'connect something with something else' and to 'establish a relationship between things'.

Key idea 2: In learning geography we are also learning the language of geography

Geography provides us with ways of seeing the world and ways of verbalising (putting thought into words). If we are to give students in school access to geographical ways of seeing and verbalising, they need to use the 'distinct literacy' of the geography, as something intrinsic to the subject. Counsell warns against bolting on literacy strategies: 'the trick is to find the natural literacy in the subject. If literacy strategies are bolted on then disaster looms' (2001, p. 14). Webster *et al.* stress the importance of literacy as an integral part of learning a subject, rather than as a set of separate skills:

'Rather than thinking of literacy as a sequence of skills to be mastered outside the curriculum, we should identify those experiences which help to form the distinct literacies of different subject areas' (1996, p. 17).

The distinct or natural literacy of geography includes far more than its specialist vocabulary. It includes the types of questions the subject asks and the way information is structured to answer these questions. It includes the concepts and organising frameworks of the subject. It includes distinctive ways of writing that have developed because of what the subject attempts to say. When we teach students geography we are also teaching them to think geographically and to use language geographically.

Key idea 3: Teachers can create a classroom 'ecology' that promotes learning through language

If we want to develop students' geographical literacies we need to create classroom environments which promote and develop the use of language in all kinds of ways.

Creating a classroom environment means paying attention to both the physical and the social environment. In terms of the physical environment, students' literacy development can be supported by:

- displays which provide stimulating things to read
- displays which celebrate students' work
- the availability of attractive reading materials – e.g. books, magazines, brochures
- dictionaries and glossaries.

Lists of words on their own do not stimulate an interest in reading and writing.

It is also important to create a supportive social environment. Webster *et al.* wrote about literacy as 'an ecology constructed in the social systems of classrooms' (1996, p. 2). 'Ecology' is a useful metaphor; geographers are aware of the complexities of ecosystems with their inter-related components and processes. To think of literacy as 'ecology' emphasises that literacy can include many elements: the processes of reading, writing and talking, the links between these, and the complexity of the relationships between teacher and learners. It emphasises literacy as a complex set of social processes as well as a set of cognitive skills. It is worth reflecting on the extent to which our classroom environments promote a wide and varied use of language (Figure 1). Students learn geography and its language through the ways they experience geography in the classroom, through the teaching and learning processes they encounter, through the language the teacher uses and presents and through the ways students are encouraged to participate in the use of language. Cairney argued that the way teachers structure classrooms 'has a strong influence on the forms of literacy that are valued' and that the patterns of social interaction and activities can limit 'the types of knowledge and literacy constructed' (1995, p. 33). Students learn about what is valued in terms of literacy as much from the hidden curriculum of the ecology of the classroom as from explicit activities.

DEVELOPING THE VOCABULARY OF GEOGRAPHY

The language of geography is not the language students use in the school grounds or at home or in their leisure time. There are many words which they encounter and use only in the geography classroom. To a certain extent, the language of geography is a foreign language and the geography classroom is a foreign country. When students study English and modern foreign languages in school, they learn to listen to, speak, read and write the languages. Detailed attention is paid to all four processes of using

Figure 1: *Reflecting on the literacy environment of the classroom.*

Reading

What are the purposes of reading in geography?

What opportunities are there for students to read in the classroom?

What provision is there for students with different levels of achievement in reading to read appropriate and challenging texts?

What do students read?

How do they read (individually, in groups, as a class)?

For how long do they read?

Are there opportunities for extended reading a) in class and b) for homework?

How are students helped to make sense of what they read?

Are students helped in their reading by written activities (e.g. DARTs) or by discussion?

Do students become aware of a wide variety of reading matter relevant to geography?

Do they have choice in what they read?

Does the physical environment of the classroom value reading?

Does the social environment of the classroom encourage reading for meaning?

Are students encouraged to read critically, with a questioning attitude towards what they read?

Are there gender differences in students' reading?

Writing

What are the purposes of written work in geography?

What opportunities are there for students to write?

Are writing opportunities differentiated?

What different kinds of writing do students do in the classroom?

Do students have opportunities to plan their writing in pairs or in small groups?

Do they have a choice in what to write in terms of topic or type of writing?

For how long do they write?

How are students supported in their writing through: (a) whole-class discussion (b) small-group discussion (c) reading?

How are students supported in their writing (a) at word level (b) at sentence level and (c) at meaning level?

Do students learn to write for a variety of audiences (teacher/imagined/real)?

Is further use made of students' written work after it has been completed?

Does the physical environment of the classroom value students' writing?

Does the social environment of the classroom encourage positive attitudes towards writing?

To what extent is writing used to help students sort out their ideas?

Are students provided with opportunities to write reflectively about what they find difficult to understand?

If written work is to be assessed, are students aware of the assessment criteria?

Are there gender differences in students' writing?

Speaking and listening

What are the purposes of classroom talk in geography?

What proportion of classroom talk is teacher talk and what proportion is student talk?

Are ground rules established for classroom talk in geography?

To what extent do students have opportunities to contribute to class discussion?

Do students have opportunities to explore their ideas tentatively in small group discussion?

Do students have opportunities to present information and ideas orally to the rest of the class?

Does the classroom environment value speaking and listening as a means of learning?

In what ways are speaking and listening related to reading and writing?

Are there gender differences in the ways students contribute to whole-class and small-group discussion?

Note: Each of these questions could form the focus for classroom-based action research.

the language. Geography teachers could learn from this. Do we encourage students to develop all four language processes, do we encourage them to listen to, read, write and speak the vocabulary of geography? The vocabulary of geography includes:

1. specialist geographical vocabulary

2. words used in everyday life, but which have special geographical meanings, e.g. relief

3. general words which are used in textbooks but are rarely encountered by students in their everyday lives, e.g. whereas.

Figure 2. Ways of supporting the development of vocabulary.

1. Identify, in schemes of work, new vocabulary likely to be encountered during enquiry work.

2. Plan opportunities for students to hear, speak, read and write new vocabulary during enquiry work.

3. Make use of vocabulary already known by asking students for ideas of words to use.

4. Draw attention to new vocabulary to be encountered in resources used for enquiry work, such as texts and videos.

5. Discuss the meaning of new vocabulary, relating it to similar words and relating it to what students already know.

6. Draw attention to the spelling of new vocabulary.

7. Use lower case letters on worksheets, displays, and on vocabulary presented on the board, on a transparency or on the computer (except when capital letters are grammatically appropriate). Students learn to read words not only by reading the letters but also by recognising the shape in lower case letters.

8. Encourage students to use dictionaries for themselves and to refer to word banks or glossaries provided.

9. Encourage students to use new vocabulary themselves orally, by devising discussion activities in which students need to use the vocabulary.

10. Encourage students to use new vocabulary in their written work.

11. Use new vocabulary in context when discussing work with individuals and groups.

12. Reinforce the use of new vocabulary by using it frequently in subsequent lessons.

13. Use new vocabulary in context in displays.

When developing the vocabulary of geography we need to pay as much attention to the second and third categories above as the first. Figure 2 provides guidelines on how the vocabulary can be developed as an integral part of learning geography. The emphasis in the guidelines is on the use of vocabulary in meaningful contexts and through repeated use by both teachers and students. Students need to develop confidence in using the language of geography. They will learn to use words by hearing them, reading them, saying them and writing them. This is what we should be giving them opportunities to do.

SUMMARY

Literacy across the Curriculum, like its predecessor Language across the Curriculum, should not be seen as an optional extra for geography teachers, providing occasional activities for some lessons. The use of language is an integral part of the learning of geography. We communicate through the use of language and we develop our thoughts through the use of language. We learn geography through the use of the language of geography. This is best learnt in purposeful contexts in classrooms in which language development is valued.

The next three chapters focus separately on developing reading, writing, speaking and listening activities and how these can contribute to the learning of geography.

'The overarching aim of
schools ought to be the
production of confident,
ambitious and critical readers
who see reading like other
aspects of their language
facility, as a key part of their
engagement in and
understanding of the world'

(Traves, 1994, p. 97).

INTRODUCTION

Much of the data that can be used for geographical enquiry comes in the form of the printed word. In their everyday lives students come across all kinds of reading matter which contribute to their own personal geographies: newspapers, magazines, advertisements, cartoons, brochures, noticeboards, CEEFAX, Teletext, text on the internet, postcards, letters, fiction and songs. In addition, they will encounter more formal 'information texts' in geography lessons, in textbooks and resources written especially for geography. All these types of text, because they contribute to how we understand the world, can be relevant to geographical enquiry.

Although the focus of this chapter is on the use of the printed word in geographical enquiry, it is recognised that most reading activities also include the use of writing and talking. The chapter outlines research findings related to reading. It then provides examples of activities to encourage the development of reading skills as an integral part of geographical enquiry.

RESEARCH FINDINGS

The research findings on reading in geography classrooms suggest that generally students have little opportunity for sustained reading. The Effective Use of Reading Project (Lunzer and Gardner, 1979) found that the kinds of reading taking place in humanities lessons in the lower years of secondary schools included students reading textbooks, reading the blackboard and reading what they had written in their exercise books. On average, however, students spent less than 15 seconds at a time reading. The project thought that such short-burst episodes of reading were unlikely to lead to the development of understanding. More recent studies have confirmed these findings. Webster *et al.* (1996) found that episodes of reading in secondary schools were likely to be brief and discontinuous: students were not given opportunities to read as part of the learning process.

Typical classroom reading activities involve answering comprehension questions which encourage a 'picking at' the text to find particular bits of information, an activity that can be done without necessarily making any sense of the text as a whole. This is well illustrated in 'Reading without having to think' opposite – an extract from a teacher's account of his action research. Although the extract is related to science, similar comprehension exercises are used in geography. We must consider whether students can answer comprehension questions correctly without having to think and without the information 'even passing through the brain on the way'. We need to question the extent to which such questions support geographical enquiry and the development of understanding.

The Effective Use of Reading Project was concerned that students should understand the meaning of texts as a whole, and devised a range of activities that focused on the structure and meaning of texts. The project activities were called Directed Activities Related to Text (DARTs). DARTs encouraged students to read and then re-read texts closely to help them understand the meaning, but also to help them understand the ways in which texts were structured for different purposes.

The project devised two types of DARTs:

1. Reconstruction DARTs in which the text was altered in some way, e.g. cut up into segments, and had to be reconstructed by the students.

2. Analysis and Reconstruction DARTs in which the text was presented as a whole. The activities were devised to enable students to:

   • analyse the component parts of a text and then
   • reconstruct the component parts into a simpler form.

Reading without having to think.

The following extract was written by Pat Jones as part of his research for the Wiltshire Learning about Learning group. He wrote it after finding, as new acting deputy head teacher, that he was expected to teach some science.

'The science course was a thoroughly prepared one, for 'less able' students, and with a bit of mugging up I was able to keep at least one worksheet ahead of the class. The first sheet was on yeast and baking. Side one gave some simple facts about the topic: "Most plants contain a green colouring called **chlorophyll**. This chlorophyll enables them to *make their own food by* **photosynthesis**." It went on to note that yeast gives *off* **alcohol** *when it respires*.

Teacher talked it through. Class turned it over and got on with the questions on the back. Jane was about third in the queue to get her answers checked. Sure enough, she had the correct answers throughout. Her correct answer to question number 6 showed that she had obviously absorbed the fact that the brewer uses yeast because it can produce alcohol. Teacher felt a bit of an automaton doing all this ticking and decided to have a quick chat.

"Why does the brewer use yeast then Jane?"

Jane panicked, looked wildly round, went red, blurted out with the unsure, rising tone of a question:

"Because it turns the beer brown, Sir?"

Teacher's turn to panic. I had witnessed the phenomenon known to all worksheet designers that a "right" answer can pass from one side of a worksheet to another without being understood or without even passing through the brain on the way...

I took this up with an experienced colleague in the science department who acknowledged the weaknesses of the approach but who told me that the main aim of this first sheet was to give them some confidence and success. They almost all got it right, almost always. It seems to me that the real lesson learned by Jane and others is that you can get right through a science topic without even having to think, and that is one quality that all teachers should seek to foster as the highest priority, the ability to think in all the many and varied ways that the brain is capable of. As the language across the curriculum debate continued and deepened, the more it impinged on the area of thinking in schools. We can't just pass down information, tell them to absorb it raw, ready for regurgitation in a later test, and expect them to learn anything. To learn something, students need to take that piece of information and build it into their own picture of the world. The information needs to pass into and around their consciousness. Only when it becomes part of the pattern in their heads does it become theirs. That means that learning needs to be a more involved and committing experience than transcribing words from blackboard to exercise book, or from one side of a worksheet to another.'

Source: Jones, 1981, pp. 5-10.

RECONSTRUCTION DARTs

The text for use in reconstruction DARTs cannot be taken directly from a textbook or newspaper article, it needs to be altered by the teacher in some way so that students can reconstruct it. This might mean cutting up text into sections and putting it into envelopes. If the sets of text are to be used again it helps if:

- they are printed on card
- they are laminated
- each set can be easily identified, e.g. by a symbol or by use of coloured paper or card, so that bits of texts from separate envelopes do not get muddled.

Two types of reconstruction DARTs are useful in geographical enquiry: sequencing and diagram completion.

Sequencing DARTs

Sequencing DARTs are suitable for types of text which describe processes and change over time. For example, sequencing DARTs could be used for an event such as the eruption of Mount St Helens which took place over a matter of weeks, or could be used for a process of change which took place over many years, e.g. the growth of a city, or could deal with changes in a landscape which took place over a longer geological timescale. Sequencing DARTs are also suitable for investigations into commodity chains, the sequential activities involved in commodity production.

For this strategy, the teacher selects or writes a text describing a sequence. The text is divided into segments, cut up and put into envelopes, with one envelope for each pair of students. The task for the students is to put the text into an appropriate sequence.

A good geographical variation on this activity is to include diagrams, maps, graphs or pictures which illustrate each stage of the sequence. These too are presented to students on separate pieces of paper. The task of the student is now two-fold:

1. to match the diagrams/maps/pictures/graphs to the pieces of text. This can require close reading and re-reading of the text.

2. to arrange the matched text and illustrations in sequence.

The activity encourages students to read text more closely than they are required to do in most comprehension exercises. It also encourages them to make sense of the text as a whole. Students often find sequencing DARTs more demanding than they appear at first sight. Examples 1a and b, Chapter 10 (pages 130-132) shows a sequencing DART related to the formation of the East African Rift Valley. In this case, students were presented with the diagrams already in sequence. The DART activity was to match the descriptions with the diagrams so that the sequence made sense. This activity was a small part of a lesson which is outlined in full in Example 1, Chapter 10 (pages 130-132).

Diagram completion DARTs

Diagram completion DARTs are suitable for types of text which describe the structure of features and their component parts. The activity focuses attention on the terminology related to different parts of the structure. A diagram completion DART could be used for investigating, for example, the water cycle, an ecosystem, a volcano or an urban area.

For this strategy, the teacher selects a piece of text and a related diagram, map or graph. The text is presented to students as a whole. The diagram is altered by removing most of its labelling. The task for the student is to:

• read the text, identifying the different components of what is being described
• complete the labelling of the diagram using the text to find the correct terminology.

This activity requires both close reading of the text and close attention to the diagram/map/graph to be labelled. Although the activity is closed, in that there are correct answers, it can be quite demanding and teachers need to experiment with which parts of the labelling are left out and which are left in for support. The activity can be differentiated by varying the amount of labelling left on the diagram.

ANALYSIS AND RECONSTRUCTION DARTs

In analysis and reconstruction DARTs the text is presented to students as a whole, although it might have been adapted from another text. There are two stages in the activity and it is important that students are involved in both stages:

1. analysis of the text, which can take the form of underlining, dividing into segments or labelling paragraphs

2. reconstruction of the text, which can take the form of lists, tables, flow diagrams, annotated maps or other diagrams.

Underlining or highlighting followed by reconstruction

This type of DART is suitable for texts in which information can be categorised in some way. The following text types identified in the Literacy Across the Curriculum framework are usually structured in some way by categories:

- a report of an event (recount text)
- an explanation of a process (explanation text)
- a discussion of an issue (discursive or analytical texts)
- putting a point of view (persuasion text)
- combinations of these in a geographical account.

Figure 1: Categorising geographical information.

This list below provides some examples and is not intended to be comprehensive.

- economic, social and environmental factors
- economic, social and environmental effects
- causes, effects, implications
- physical causes, human causes
- short-term effects, long-term effects
- who gains? Who loses?
- advantages, disadvantages
- arguments for, arguments against
- facts, opinions
- female roles, male roles
- statistics, text related to statistics

The kinds of categories commonly used in geography texts and which can be used as the basis of underlining in DARTs are shown in Figure 1.

For this strategy, the teacher chooses a well-written piece of text which is structured to a certain extent by implicit categories. The teacher determines the categories that would be useful in analysing the text and provides these in the instructions to the students. The text is presented as a whole but needs to be in a form that students can mark, e.g. reproduced on a worksheet or presented on a computer screen. The task for the student is to:

- identify the different categories
- distinguish between them by marking them differently (different types of underlining, different highlighting pens, or by use of underlining and italics on a computer)
- reconstruct the information gained from the analysis in a different form, i.e. in tables, diagrams, (e.g. Venn diagrams), and flow charts, or concept maps. These would represent the text as a whole and would reveal the implicit structure of the text.

This activity requires close analysis of the whole text, analysing it into its component parts. The reconstruction of the text includes only the essential information from the text, but this is now structured in a meaningful way to produce what is, in fact, a form of notes. This activity, as well as enabling students to analyse the text presented to them, introduces them to skills of note-taking which can be used in other contexts, e.g. when they have printed out information from the internet and need to make sense of it or when they are summarising information presented in a textbook or newspaper article. Example 7e in Chapter 13 (page 178) shows a DART activity in which students analysed positive and negative aspects of the lives of street children in Nairobi through the use of two different kinds of underlining. In this example, students reconstructed the information in two ways. First, they put the information they had analysed into a simple table. Second, they used the information they had to write a commentary for a charity video (Example 7f, Chapter 13, page 178). This example extends the original DARTs ideas by including a short piece of writing to give students more scope to make sense of the information for themselves and suggests a useful modified DARTs sequence which incorporates extended writing as follows:

1. analysis, encouraging close reading

2. reconstruction in note form, encouraging identification of key points

3. extended writing, encouraging further thinking about the ideas presented in the text.

The activity made further demands on students' literacy skills; as the text was written by a Kenyan girl (whose first language was not English), the first task for the students was to correct the English.

Text labelling followed by reconstruction

This type of DART is suitable for the complete range of text types and for text combining different types. For this strategy the teacher selects or devises text with well-structured paragraphs. The text is presented as a whole to students, but can be altered to provide space for labelling above each paragraph. The text needs to be presented on paper or on a computer screen so that students can add to it. The task for the student is to:

1. devise a suitable sub-heading for each paragraph

2. devise a suitable heading for the text as a whole

3. reconstruct the sub-headings into a diagram or table.

This activity requires close reading of each paragraph and develops the skill of summarising meaning. A variation on this type of DART requires students to select photographs, from a given collection, to match each paragraph. This can most easily be done on computer. In Example 1, students, working at computers, were provided with a set of paragraphs providing information about the floods in Mozambique (in *Microsoft Publisher*) together with a folder containing 40 photographs. In order to match the photographs to the text appropriately, students needed to read the text closely for meaning and also to scrutinise what was shown in the photographs. They then had to place the paragraphs and photographs in chronological order.

Example 1: DART activity: Picture Editor. Source: Giles Hopkirk.

In this activity, students were presented with the following paragraphs, already set out in *Microsoft Publisher*. At the side of each paragraph was a space for a photograph. The task for students was to select an appropriate photograph from a collection of 40 provided in a folder on the computer and to place it next to the paragraph.

1. 10 February: The government of Mozambique appeals for international help. 11 February: Specialists arrive to assess the disaster. 21 February: Cyclone Eline brings more rain. Over one million people start the trek to higher ground.

2. Things get even worse. People are forced onto smaller and smaller patches of dry ground. Many are swept away.

3. Helicopters from the South African military abandon plans to drop food and concentrate on evacuating people. Western media start to get pictures of the flood from the helicopter crews.

4. Torrential rains started flooding the south of Mozambique in early February 2000, but it wasn't until the end of the month, especially the weekend of 27-28 February, that western media began to realise how serious the situation was.

5. 7 February: Southern Mozambique has the heaviest rains for 50 years. The Zambezi, Limpopo, Save and Incomati Rivers burst their banks or are close to doing so.

6. Heavy rain falls over the neighbouring countries of South Africa, Zambia and Zimbabwe. Fearing that dams will collapse under the extra weight of water in the reservoirs, the authorities open the floodgates sending walls of water into already flooded areas.

7. The tiny fleet of helicopters spends a frantic week trying to rescue people at risk of drowning in the rising water. Dramatic pictures and stories arrive every day from journalists and photographers, including the birth of a baby girl in a tree, assisted by a medic from one of the crews.

8. The US increases its aid from US$600,000 to US$6.5 million. Britain sends four RAF Puma helicopters. Attention focuses on children separated from their families, the threat of disease and rebuilding the country.

INCORPORATING DARTs INTO GEOGRAPHICAL ENQUIRY

Directed Activities Related to Text are, as their name indicates, activities which are controlled by a teacher. In the use of DARTs, teachers make many of the decisions related to the enquiry; they choose the focus of study, select appropriate data, and decide how it is to be used to explore the issue being investigated. The limitations of DARTs need to be recognised. They are usually closed activities, with right answers. They do not give students opportunities to identify questions, to suggest sequences of investigation or to look for sources of information. Also, although they can involve students in concentrated analytical work, DARTs activities are not in themselves necessarily very exciting. They need to be made purposeful by the contexts in which they are used, with particular attention taken to 'creating a need to know'. Also DARTs activities present text as authoritative data, rather than data to be questioned in terms of its reliability, its origins, its purposes and its points of view. If students are to become critical readers of text, then the debriefing following the use of DARTs needs to include critical questions.

Figure 2: A procedure for incorporating DARTs activities into geographical enquiry work.

Creating a need to know

- The teacher uses a stimulus of some sort to create an interest in the focus of the enquiry.
- The teacher elicits what students already know about the focus of the enquiry.
- The teacher establishes the focus of the enquiry, generates curiosity, and discusses the questions that frame the enquiry.

Using the data

- The teacher provides students with copies of the text, as evidence to be scrutinised. The teacher reads the text to the class slowly, ensuring that all vocabulary is understood.
- The teacher gives instructions on how to carry out the DART activity (preferably both orally and in writing) and explains how this will contribute to the enquiry.
- Students work in pairs on the DART activity using a variety of skills (sequencing, sorting, underlining, highlighting, labelling photographs).
- (A possibility, if it seems appropriate.) Students discuss in an interim whole-class discussion for a few minutes what they have already done with the data: examples of matching of pictures and text, examples of what underlined, examples of headings.

Making sense

- Students work in pairs completing the DART activity, reconstructing the information (into diagrams, tables, flow charts).
- While the students are doing the DART activity, the teacher provides support through interactive discussion with pairs of students. The teacher makes a mental note of the ways in which students are reconstructing the text so that this can be used in the debriefing session.
- The teacher debriefs the activity asking questions which probe the meaning and the value of the text which could include the following:

  1. What did you underline or highlight? (Collecting examples of what students did to analyse the text.)
  2. What did you show on your diagram/flow chart/table? (Collecting examples of what students did to reconstruct the text.)
  3. What have we found out to help us answer the questions we started with?

Reflecting on learning

- The teacher continues the debriefing activity with questions which evaluate the enquiry and which consider the thinking involved.

  1. How reliable is this text as evidence? Who wrote it and why? Is it biased?
  2. Do you think other texts would tell a different story?
  3. What other information might we need to answer the questions in more detail?
  4. Where would we find this?
  5. What skills did you use?
  6. Could you use these skills when making notes for other enquiry work?

- Students could be given a similar text to analyse and reconstruct for homework so that they can apply the skills they have learnt in a new context.

Figure 2 suggests a classroom procedure for incorporating the use of DARTs into geographical enquiry. Used carefully, DARTs have a contribution to make. They encourage the development of reading skills which are necessary for enquiry work and which are rarely developed in the geography classroom. They focus on understanding rather than on separate bits of information. They encourage students to analyse text closely and to make sense of it. Students can learn transferable skills of note-making through the use of DARTs; they can learn how to reconstruct the meaning of text in simplified forms. Students can then use these skills when they need to use the printed word as data in more open geographical enquiry work, including text collected from the internet.

The central idea of DARTs, i.e. developing activities which focus on the meaning of text, can be applied to more open activities. Mysteries, for example, are similar to reconstruction DARTS in that they present short pieces of text for students to study closely. They differ from DARTs as developed by the Effective Use of Reading Project in that there is no one correct way of sorting the data and students make their own decisions about how to categorise it. This can make students more involved in the process of making sense of the data (Leat, 1998; Leat and Nichols, 1999).

Example 2, Chapter 10 (pages 133-134) shows an activity in which students categorise fragments of text which is more open than most DARTs activity, but which is not set in the context of a mystery.

USING LIBRARIES AND COMPUTERS AS SOURCES OF DATA

Teachers need to use strategies other than DARTs to enable students to develop the skills of locating resources, finding information in the located resources, and selecting what is relevant for a particular investigation. In DARTs all these choices have already been made. In most cases, these choices have also been made in geography textbooks. Students need some opportunities to locate information for themselves and to search for relevant data and to become aware of their criteria for selecting data. Libraries, CD-Roms and the internet can all be used to develop these skills. If, however, students are to make good use of the vast amount of information available, then they need some guidance. These skills, like those associated with DARTs, are best developed in purposeful contexts. Figure 3 offers a procedure for incorporating the use of library skills and the use of the internet into enquiry work. The activity in Example 2a was devised to introduce year 7 students to the school library, to the use of atlases and to some aspects of the enquiry process. The task for students was to plan an imagined journey from the North Pole to the South Pole, a task to be completed during four lessons in the library and some homework time. One feature of the activity was that the students could exercise some choice. They could choose which line of longitude to follow, which countries on that line to investigate, which resources to use and how to present their findings. One possibility for reporting their findings was through imagined e-mails, using a writing frame for support (Example 2b). Before they started their imagined journeys, students were provided with information about how the work would be assessed (Example 2c). It is worth considering how the kinds of choices incorporated in this example could be included in other enquiry work.

Figure 3: Incorporating the use of library, CD-Roms, and the internet into enquiry work.

Creating a need to know

- The teacher provides a stimulus of some sort to create an interest in the focus of the enquiry.

- The teacher elicits what students already know about the focus of the enquiry.

- The teacher establishes the focus of the enquiry, generates curiosity, and discusses the questions that frame the enquiry.

- The criteria by which the work will be assessed might be negotiated.

Using the data

- The teacher provides some guidance on how to use the library or computer (preferably both orally and in writing). Note: It might be wise to limit the search by providing students with details of some resources (e.g. titles of books, encyclopaedias) or by making an initial selection (by providing a table of pre-selected books, by providing a limited number of websites, or by downloading a limited number of pages from the world wide web onto the school intranet).

- Students locate relevant resources.

- Students select relevant information and make notes, or cut and paste it for future use.

Making sense

- Students use their selected information to make sense of the topic for themselves. This could include planning how to present the information orally to the class, visually for display or in written form for a report.

- Students are guided by teachers through interactive discussion about how best to present their work.

- Students present their work.

- The work is assessed in some way, either by the teacher or by other students.

Reflecting on learning

The teacher debriefs the activity by asking questions which probe *what* has been learnt. Debriefing questions might include the following (adapted for the particular context):

1. What did you learn from this work? What was the most interesting/surprising thing you learnt?

2. Has this work made you change your views on this topic/place? In what ways?

3. How would you explain what you have learnt to someone younger than you?

The teacher debriefs the activity by asking questions which probe *how* the learning has taken place.

1. Which resources did you use?

2. Which resources did you find most useful for your investigation and why?

3. Which resources did you find least useful?

4. Do you think these resources were reliable? In what ways?

5. I wonder if you would have found out something different from other resources?

6. How did you set about combining information from different sources?

7. What did you find easy? What did you find difficult?

8. What advice would you give to another class doing the same investigation?

Another example of supported library research is provided in Example 3 in which year 9 students investigated hurricanes. Students were supported by:

- the provision of a box of selected books about hurricanes
- a list of library locations
- a list of questions to structure the enquiry

For those who needed additional support, a writing frame was provided with the following sentence starters:

- A hurricane is …
- Hurricanes happen because …
- Hurricanes happen in these places …
- Hurricanes can affect people by …
- Overall, I can say this about hurricanes …

Example 2: Pole to Pole activities. Source: Geography department, King Edward VII School, Sheffield.

(a) Library research

Pole to Pole – Y7 assessed work

Imagine you are making a journey from the North Pole to the South Pole. You have to choose a line of longitude to travel along. It must be a line that passes through at least three countries. Use the lesson in the library to find out about the countries you will be travelling through. You could also find out information from CD-Roms, the internet, travel brochures and use your own experiences on holiday.

You must then write about your journey, about the places you saw, the people you met, the adventures you had. You could write a diary, you could write e-mails sent from internet cafés on your route. You could send home a postcard from each country or do a tape of your adventures. You could collect souvenirs on your travels and include them with your work. You *must* include a map of your route and perhaps maps of the countries you visited.

This project is to be done on paper, put in a folder and handed in on .... You will be marked on your map skills, the research you have done, the information you found out, your understanding of how life is different in other countries, and your writing skills.

Have a good trip!

(b) E-mail writing frame

E-mail

| From: |
| To: |
| Sent: |
| Subject: |

Hi there!

I'm having a very exciting journey from Pole to Pole. I have now arrived in _____

_____

The places I have seen are _____

_____

_____

I have met lots of interesting people. Yesterday I met a family in _____

_____

(c) Assessment of library research

Pole to Pole – Y7 assessed work

Name ....

Form ....

Skills (5) Marks for map work, organisation, presentation, pictures, diagrams

Research (5) Marks for finding out information and using the library

Knowledge (5) Marks for information about countries, both human and physical

Understanding (5) Marks for understanding how life is different from or similar to life in the UK

Writing (5) Marks for the quality of the written work

Total (25)

Effort grade

Example 3: Library research: hurricanes. Source: Richard Davies.

Instructions

- Today we are going to do some research in the library.
- You will then write up this research as your homework to produce a small project on hurricanes.

Where do I get the information?

The information can come from the following sources:

The Hurricanes book box

The following locations in the library:

- 551.5
- 551.6
- 363.3
- 363.4
- Encyclopaedias
- Geography textbooks in the geography section

What do I do with the information?

You write notes in the back of your book, draw sketches and diagrams to help you with your project. Make sure that next to your notes and diagrams you write where you got your information.

What does the hurricanes mini-project have to include?

You have to answer these questions:

- What are hurricanes?
- Why do hurricanes happen?
- Where do hurricanes happen?
- What effects can hurricanes have?
- Extra: what do people do to cope with hurricanes?

Example 4 provides instructions for higher achieving year 8 students to carry out their own investigation of oil using both the internet and the school library as sources of information. Students had three lessons to carry out the activity. They were supported in their work by the provision of:

- a list of key questions
- a list of useful websites
- instructions for using the computer
- resources in the school library and geography departments.

Although the topic was chosen for the students, they were encouraged to ask additional questions, to decide whether to work individually or in groups, to decide how to present their findings and to negotiate the details of how to carry out the enquiry.

OTHER WAYS OF INCORPORATING READING INTO GEOGRAPHY

There is a wide variety of engaging reading matter that can stimulate interest in geography, including accounts of travel, fiction, poetry, travel books, magazines and newspapers. Whether students are aware of this reading matter and whether they make use of it depends to some extent on encouragement given by the teacher. This can be done in a variety of ways:

- Classroom displays can include thought-provoking extracts or quotations from books (fiction and non-fiction), newspapers, brochures and magazines.

- Teachers can read out loud to students from interesting texts, including fiction, poetry, songs, non-fiction, topical letters to newspapers, newspaper articles, etc. This can provide a stimulus for the start of a lesson. For example, the first chapter of *Refugee Boy* (Zephaniah, 2001) would provide a thought-provoking start to an investigation of refugees. Another useful extract for teaching about migration from Mexico to the USA can be found in *Full Circle* (Palin, 1997).

- In these readings, teachers could try to include voices rarely heard in the geography classroom, text written by different people from all over the world, from different perspectives. The internet gives teachers access to a far greater range of voices than was available even a few years ago. Real voices

Year 8 Energy and resources research project

Task sheet: Project: oil

Note: This is a very important task sheet that provides information for at least three lessons. Keep it safe!

The project

- You are to produce a project on the topic of oil.
- Try to think of what this word means to you and the type of information you need to find.
- Listed below are questions you should attempt to answer. The project will be based mainly on research in the Learning Resources Centre and the internet and will involve using computers to present the information you find.
  - Background information on what oil is and how it forms.
  - Where in the world do we find oil?
  - Why is oil so important? What are its many uses?
  - Will oil run out? If so, when?
  - Where are current explorations for new oil reserves taking place?
  - How can oil harm the environment? Examples of oil spill disasters are needed.
  - How can oil spill disasters be prevented or their environmental impact reduced?

These are just some suggestions, you may have others of your own. Try to think of ways to expand the project and gain extra marks.

What do I do now?

You may be working in groups to complete this project. You must aim to work well together and share the workload any time spent at a computer evenly between you. How well you work together will form one part of the assessment for this work.

How am I going to present this work?

You have a number of choices:

1. As an official document printed on A4 paper. Think about the style of presentation, the use of graphics, images and text. Use headings for chapters, and sub-headings.

2. As a wall display. Mounting your print-outs on backing paper and producing a display with a range of text and graphics and images.

3. As a *PowerPoint* presentation. A slide show displaying the facts researched using text, images, animations and sound.

4. In a style of your choosing. This must be agreed with your teacher.

How much time do I have?

Depending on the availability of the ICT and Library (LRC) rooms you will have up to three lessons to research and put together your presentations. There may also be follow-up lessons in normal classrooms to allow you to put the finishing touches to your projects. You also have a number of homeworks during which you are expected to contribute a great deal to these projects.

When is the deadline?

You will now agree a deadline with your teacher for when to hand in your work. Failure to meet this deadline will result in a departmental detention. If you have any problems or questions make sure you see your teacher in plenty of time before the deadline.

It is your responsibility to print off and hand in this work, in the presentation style agreed with your teacher, by the deadline agreed below.

Project deadline ....

The research

Use the internet to find the information required in order to answer the main project questions and to provide any other interesting facts about oil.

The internet is not the only source of information, you may also use textbooks and CD-Roms from both the geography department and the LRC.

A number of websites are provided below to help you start. These can be found through the geography department website on the intranet. Do not forget to add these to your 'favourites' list.

Instructions for using the computer

1. Log on and select I-Net from the topic selector.

2. Open a *Word* document. This creates a space into which you can copy and paste and edit the information you find on the internet. Now minimise the *Word* document so that it appears as a small icon on the toolbar at the foot of the screen.

3. Click on the BCHS Intranet followed by subjects and

geography. Click on one of the above websites and once connected add it to your favourites to save time in future.

4. Begin your web investigation, copying and pasting relevant information into your *Word* document. Remember to copy a picture or a graphic, use the right hand button on the mouse to click on the object and then select copy from the menu.

5. Do not forget to save your *Word* document regularly and to add good websites to your favourites folder so that you can locate them again quickly and easily for use in future lessons.

Do not copy and paste everything you see. Read it and be selective. Is it relevant to your case study? Do you understand what has been written? Have you reached your own conclusions?

6. Make sure your learning is also enjoyable! Do any of the websites have games that you can play to help learn about this topic? If so, it may be useful to write an evaluation, or design your own game!

7. Once you have collected all the information you can, you need to begin editing your document or presentation. Think about the use of fonts, graphics, headings.

8. Use and practise a range of ICT skills and geographical knowledge to produce a high-quality, detailed, organised and well-presented document/*PowerPoint* slide show. Make sure you come to your own conclusions.

9. Hand in an excellent piece of work.

Useful websites to get you started

Oil spills: http://www.soton.ac.uk/~engenvir/ environment/water/oil.slicks.html

Oil information: http://seawifs.gsfc.nasa.gov/ OCEAN_PLANET/HTML/peril_oil_pollution.html

French oil spill: http://news2.thls.bbc.co.uk/hi/english/ world/europe/newsid%5F561000/561754.stm

Environmental perspectives on oil: http://www/greenpeace.org/

Government information on energy: http://www.dti.gov.uk/EPA/eib/

Oil reserves – excellent site: http://www.petroleum.co.uk/politics/htm

Sea Empress oil spill: http://www.swan.ac.uk/biosci/empress/

---

of real people can be used as data for enquiry, in speech bubble displays on worksheets and notice boards and as information for role-plays.

- Books related to geography can be bought for the school library or the geography classroom including: fiction, travel literature, encyclopaedias and non-fiction accounts. Students vary in what they like to read, but many students in years 7, 8 and 9 are capable of reading much longer pieces of text than they normally encounter in the geography classroom.

SUMMARY

Research suggests that insufficient attention is given to reading in most geography lessons. There is a lot of scope for developing students' ability to read for understanding, both through structured DARTs activities and through providing guidance on using libraries and the internet. Good debriefing of these activities can make students question what they read at the same time as they learn from what they read. There is also scope for making students aware of a wide range of books that can help them develop their understanding of the world. We tend to limit rather than extend what students read in geography. Instead, our aim should be to enable students to read about the world, both inside and outside the classroom, independently, critically and with understanding, interest and enjoyment.

*'Writing is only important
if there are things you
want to write about.
The primary task of a
teacher of writing for
children (and inexperienced
writers) is not to induct into
form but to help students
discover the purposes and uses
of writing: as satisfying ways
of interpreting, preserving
and sharing experience and
imagination.Writers need to
find both the "what" and
the "how" of writing at the
same time'*

(Rosen, 1994, p. 196).

INTRODUCTION

Students spend a considerable amount of time writing in geography lessons; it is part of the culture of the classroom. What is it all for and why do students find it so difficult? This chapter outlines some of the research into writing. It then discusses how students' writing can be developed throughout the process of geographical enquiry to promote understanding.

RESEARCH INTO WRITING

What kind of writing takes place?

Research suggests that what students write in secondary school lessons is often limited. In the secondary geography classroom researched by Webster *et al.* (1996) writing was restricted to copying and transferring information. They reported from their research, a personal, social and environmental education lesson in which:

'students completed a worksheet on personal characteristics, filling in spaces in sentences, completing stem sentences or responding to questions with single words. Examples of this kind could be found in almost every lesson across the curriculum and are highly familiar to secondary teachers ... there were very few occasions when teachers made explicit the processes of enquiry which are strategic to science, maths, design or the humanities, and which are supported and shaped through specific kinds of reading or writing' (Webster et al., 1996, p. 133).

The small amount of extended writing taking place generally in geography at key stage 3 could be related to the limited demands for extended writing in geography at GCSE and even at AS-level. Butt noted that:

'Geography examination papers now tend to require students to write single sentence answers or, if more marks are to be awarded for a more complex answer, space is given for a three- or four-line response. Only in the (proportionally fewer) "higher order" questions are students expected to engage in anything that approaches extended writing in the generally accepted sense' (2001, p. 17).

Research suggests that these limited forms of writing do not help promote understanding. There is considerable scope for improving writing in geography classrooms.

Writing for different audiences

When we write, we usually have in mind who is likely to read what we write and take this into account. The reader of what we write provides us with an 'audience'. The audience could be ourselves; we might write notes to help us remember things or diaries to help us reflect. If we write to other people, e.g. in a letter, we take into account what they already know and what we need to make clear. School writing is not usually like this. The concept of 'audience' for written work was developed by the team of researchers working on the Schools Council Writing Across the Curriculum Project (1971-74). Britton *et al.* (1976) categorised the types of audience that secondary school students wrote for. The project found that the most common category was 'pupil to examiner' (48.7%). In this category, the writing marked an end point of learning and was for the teacher to assess. In geography, the percentage of writing for this audience of 'teacher as examiner' was particularly high, ranging from 75% for 11-12 year olds to 98% for 15-16 year olds (Britton *et al.*, 1976, pp. 133 and 135). The project team thought that writing for the audience of 'teacher as examiner' was limited and neglected 'a vital educational function of writing' (Britton *et al.*, 1976, p. 196).

Butt (1993, 2001) carried out classroom-based research to investigate the effects of writing for audiences other than 'teacher as examiner' on students' learning in geography. He gave students writing tasks aimed at a variety of audiences that included an aid agency, a shanty town dweller, a hospital consultant, and an MP. Butt found that 'audience-centred' teaching led to an increase in work-related discussion, an increase in the number of questions asked, an appreciation of the audience's perceived viewpoints and clarification of personal values. Students who were willing to experiment did improve their learning of geography.

Taking the work of Butt and Britton *et al.* (1976) into account, it is possible to envisage four categories of audience for written work in geographical enquiry.

1. *The students themselves.* This writing would include notes made to collect information from data and reflective writing in the form of journals or evaluations.

2. *Teacher as examiner or assessor.* In this category, when students write they have no audience in mind other than the teacher who is going to mark the work. The written work could take different forms, e.g. a report of an investigation; an account describing and explaining an event; a discussion of an issue. If students are to succeed in geography, they need to learn to write in these ways, as this is the style of writing most required for public examinations. This kind of impersonal writing can help students clarify their ideas, analyse issues, develop arguments and communicate their findings.

3. *An imagined audience.* In this category, students write for an audience that they can imagine, even if it is the teacher who assesses the work. In Butt's research, referred to above, the audiences were imagined in that the writing was read only by the teacher. A list of possible imagined audiences appropriate for geographical enquiry work is shown in Figure 1. This kind of writing encourages a more imaginative engagement with the topic being studied and is more motivating for many students. To succeed in geography, students need to be familiar with this style of writing; they are often presented with letters, advertisements, etc., in GCSE and A-level geography examinations. Sometimes they are asked to write letters or design posters. The texts that are used in geography at all levels are much wider than the 'information texts' listed in the Literacy across the Curriculum folder (DfEE, 2001a). Texts used in geography related to the printed word (the term 'text' has much wider connotations in geography now) include expressive writing, literature, journalism, advertisements, promotional literature, etc. These kinds of texts can provide data for enquiry at key stage 3, and students can learn to understand them better by writing for the kinds of audiences implicit in them.

4. *A real audience.* In this category students' writing is actually going to be read by people other than the teacher. Real audiences can include the head teacher (Example 3c, Chapter 13, page 172), parents, the local community through a newsletter, children in the feeder schools, and other people in the class or school (on displays). If teachers want to make geography real and want students to engage with real issues, local, national and global, then they could encourage writing for real audiences.

Figure 1: 'Imagined' audiences for written work in geography.

Geographical information is presented in a wide variety of forms in the world outside the classroom. The following forms of writing, each aimed at particular audiences, can be used for enquiry work at key stage 3:

*   **newspapers or newsletters:** particular audiences could be identified, e.g. the school, the local community or town, readers of a local newspaper

*   **Teletext or CEEFAX news reports** (limited to 50 words a page). For topical events these can be compared with the real thing.

*   **radio or television news reports** (limited in length to a few minutes)

*   **weather forecast** (e.g. Example 3, Chapter 9, page 117)

*   **commentary for video** (e.g. Example 7f, Chapter 13, page 178)

*   **holiday brochures:** possibly targeted at different types of tourist

*   **information boards** for tourist attractions

*   **leaflets about a particular place,** e.g. Mother Shipton's Well in Yorkshire, aimed at primary school children

*   **publicity leaflets about an issue:** aimed at those who would be interested, e.g. local inhabitants

*   **letter to a Member of Parliament or town councillor**

*   **letter or postcard to a friend or relative**

*   **e-mail to a friend or relative** (Example 2b, Chapter 5, page 60)

The ground rules of writing

'There is something I
don't know

that I am supposed to know.

I don't know what it is I
don't know,

And yet am supposed to know

And I feel I look stupid if
I seem both not to know it
and not know what it is
I don't know.'

(Laing, 1971, see
www.oxford.ntc.co.uk/
'knots'.htm).

Sheeran and Barnes (1991) researched the difficulties students have in writing. They noted that even students who had 'joined in successfully in the discussion in a lesson, showing every sign of understanding' handed in written work that was 'quite inadequate'. They found that in order to produce acceptable writing, students had to become adept at reading the mind of the teacher, 'as many important instructions are left unspoken'. They referred to these tacit expectations of teachers as 'ground rules'.

Example 1 illustrates some of the problems students have with understanding the ground rules. The students quoted in this example had carried out a survey into energy use (Example 2, Chapter 12, pages 157-158). The printed task sheet suggested that the introductory paragraph should include an answer to the question 'How did you go about answering the questions?'. Kerry wrote a narrative account which included details that were irrelevant to the survey, e.g. 'our teacher went away and typed

Example 1: Extracts from Energy Survey reports.

Extract A: Kerry

> To answer our first question on what energy was used 50 years ago we all as a class thought of questions which would be suitable to ask someone over 50 on the main sources of energy used 50 years ago. The class as a whole then chose the best questions we thought were suitable and our teacher went away and typed them up to make our questionaire. Then it was our turn to go away and ask a member of our family the questionaire. We then came back to class to analyse the information and create a response. Our teacher had already typed up a chart on which to pass around the classroom and write in our questionaire responses. Most of the answers where roughly the same with only a few choosing another answer.

Sheeran and Barnes (1991) argued that students' understanding of the ground rules for writing, often specific to particular subjects, was essential for success. They thought that it was important for teachers to communicate the ground rules to students. This could not be done simply by giving clearer instructions. Teachers needed to become aware of how students were interpreting what was required, so there needed to be interactive discussion. This might include discussion of:

- the purpose of the piece of writing
- possible ways of tackling the task
- ways of structuring the writing
- examples of similar work, both good and bad
- what the students think is required from them.

Extract B: Shahid

> created a questionaire with a range of simple questions which most adults I could answer without much trouble. We

Extract C: Lee

> 2 how we went a bowtit? fers we reden some awestyon in class and there miss invert and put them into a awestyoner and we went away and then we aske someone 50 or owther 50 years old.

them up'. Lee struggled with basic skills of spelling but also wrote an account that included irrelevant information, e.g. 'and we went away'. In contrast, Shahid wrote one sentence. All the students knew the answer to the question; they knew what they had done in an everyday sense, in their everyday way of looking at the world. What they did not know was how much of this to include in their report in spite of the guidance of the question; they did not know what was relevant and what was not. The teacher did not expect the amount of detail included in Kerry's work but wanted more than what Shahid wrote, but how were they to know? The ground rules for this, like for many other writing tasks, were unspoken. When students write for the audience of teacher as examiner, they often find it difficult to decide what is relevant and what can be taken for granted, as students are aware that the teacher already knows what they are going to write about.

Only when teachers have some awareness of how students understand what is required can they set about making the ground rules explicit.

The EXEL Project

The Exeter Extending Literacy Project (EXEL) was set up by Wray and Lewis (1997). The project, based on constructivist ideas, developed a range of strategies which included the use of a sequence of activities to support reading, grids to support enquiry and writing frames to support those with writing difficulties (see 'The EXEL Project and writing frames' below). Several of the strategies incorporated in the 'Literacy Across the Curriculum' (DfEE, 2001a) strand of the National Strategy for Key Stage 3 have their origins in Wray and Lewis's work.

The EXEL Project and writing frames

The Exeter Extending Literacy Project (EXEL) was set up in 1992 by David Wray and Maureen Lewis, with the aim of enabling primary school children to use literacy more effectively as a means of learning. The project, in collaboration with teachers, developed a range of strategies to improve both reading and writing.

The project developed a model, the EXIT model (Extending Interactions with Text), based on constructivism, which identified ten processes involved in learning with texts (Wray and Lewis, 1997):

1. Elicitation of previous knowledge
2. Establishing purposes
3. Locating information
4. Adopting an appropriate strategy
5. Interacting with text
6. Monitoring understanding
7. Making a record
8. Evaluating information
9. Assisting memory
10. Communicating information

In order to provide support for these processes, Lewis and Wray developed two types of support, know, want, learn

(KWL) grids and writing frames. The KWL grid consists of three columns, headed by the questions:

- What do I *know?*
- What do I *want* to find out?
- What did I *learn?*

The EXEL Project found that these grids helped children see the stages of their research and provided them with a logical structure. The teachers working on the project inserted an additional column with the heading 'Where can I *find* the information?' They termed the four-column grids KWFL grids.

Writing frames were devised by Wray and Lewis 'to scaffold and prompt children's non-fiction writing' and were found to be 'especially useful for children with reading problems' (1997, p. 27). The writing frames were produced for different types of text and consisted of a skeleton outline together with sentence starters and connective words appropriate to that text type. The writing frames were intended to be used as part of a four-stage process of learning, based on the kinds of teacher support suggested in Vygotsky's zone of proximal development (see pages 28-30):

The EXEL Project and writing frames ... continued

1. **Demonstration**, in which the teacher models what the children are expected to do, preferably with a running commentary by the teacher thinking out loud about the activity.

2. **Collaborative activity**, in which the teacher and children work together, with the teacher possibly taking responsibility for the difficult parts, while the children carry out the easy parts.

3. **Supported activity** in which some form of scaffolding is provided, such as grids and writing frames and support through dialogue.

4. **Individual activity**, in which the child has responsibility for carrying out the activity without support.

The aim was that, following the use of the writing frame, children would be able to write in a similar form, without needing its support. Like all scaffolding, the writing frame was intended to provide temporary support only.

Figure 2: A framework for learning through enquiry: focus on writing.

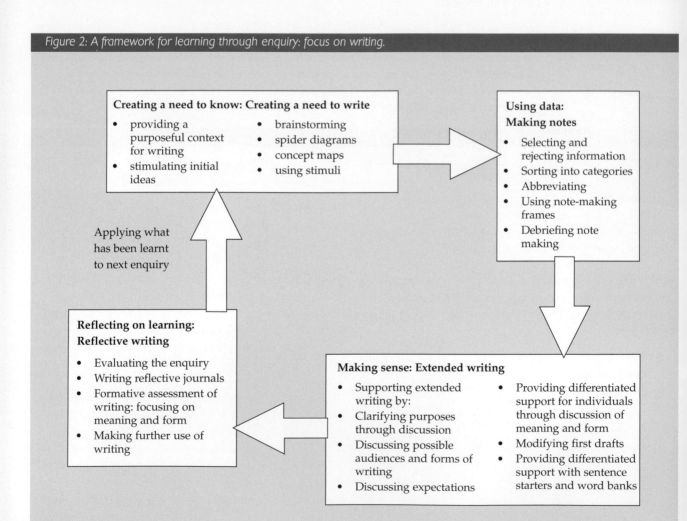

USING WRITING TO SUPPORT LEARNING THROUGH GEOGRAPHICAL ENQUIRY

Writing can contribute to learning throughout the enquiry process. It is possible to think of developing writing in four stages (Figure 2) related to the four essential aspects of enquiry identified in Chapter 3:

- needing to know: getting started with writing
- using data: making notes
- making sense: extended writing
- reflecting on learning.

NEEDING TO KNOW: GETTING STARTED WITH WRITING

Purposes and strategies

Research into the process of writing has emphasised the importance of initial activities - activities which get students started on the 'journey' of writing (D'Arcy, 1989). Such activities have several purposes. They enable students to recognise and value what they already know and to make connections between this and what they are about to study. They enable students to generate and share ideas. They enable teachers to find out what students already know, how they think and what misconceptions they might have. Teachers need to get inside students' minds and know their starting points of learning if they are to support them well. Starter activities for initiating ideas for writing include:

- providing a purposeful context
- thought showers
- spider diagrams
- concept mapping
- frameworks to prompt thinking
- using stimuli.

All of these strategies can be used to create a 'need to know'. These strategies are not new; they are based on research and development across a range of subjects by many researchers, including Wray and Lewis (1997), D'Arcy (1989), Ghaye and Robinson (1989), and Leat and Chandler (1996), and Nichols with Kinninment (2001).

Providing a purposeful context

If the outcome of enquiry work is to be a piece of extended writing, students need to be aware of this at the outset so that the starter activities are focused on what is eventually required. The nature of the initial jottings and of the notes taken will depend on what type of extended writing is eventually needed. This sense of purpose can be illustrated with examples of two lessons given by PGCE student teachers and used with year 8 students. In one lesson the writing task was for students to write a commentary for a four-minute section of a video on rivers. This purpose was integrated into the whole lesson. The room was set up as a BBC recording studio. The task was introduced with the teacher reading out a desperate fax (imagined of course) from the BBC about a missing four minutes of commentary. They needed a commentary by 3.00p.m. This scenario created a purposeful context, a need to know, an audience for the writing and criteria for note taking. This provided support throughout the lesson. The starter activity, the reading of the fax, established the purposeful context.

In a contrasting lesson, a PGCE student showed a video of flooding in Bangladesh. Year 9 students then answered questions about Bangladesh from a textbook. Five minutes before the end of the lesson the student teacher presented the homework task: to write a newspaper account of the floods in Bangladesh. Although the data in the lesson provided the year 9 students with appropriate information, none of the activities had prepared them for writing a newspaper report. This lesson could have been set in a purposeful context. The students could have been in role as journalists from the start of the lesson, collecting information for their newspaper from the video and supplementing this with information from the book with this purpose in mind. There could have been discussion on what might be included in a good newspaper account. The starter activity could have provided students with details of their flight to Bangladesh for their fact-finding mission.

Thought showers

The main purpose of using a thought shower (sometimes called 'brainstorming' sessions) is to elicit existing knowledge. This enables teachers to get into students' minds, to become aware of their knowledge, their understandings, their misconceptions and their opinions. Thought showers enable students to make links between what they already know and what they are about to study.

Using a thought shower can be managed in different ways. It can be done orally, which has the advantage of being spontaneous and quick. Alternatively, students can write down their ideas, either on one piece of paper, or on separate cards for later sorting. Writing down ideas has the advantage of involving all students in thinking. Thought showers can be created individually, encouraging all students to think for themselves, or they can be done by pairs and groups which has the advantage of generating more ideas. Teachers have to decide which approach suits a particular situation. A procedure for developing thought showers is shown in Figure 3.

Figure 3: Procedure for using thought showers.

Stage 1: Bringing ideas to the surface

Students are asked to think of lists, questions, ideas, memories or vocabulary in response to questions such as:

- What are the first five things that come into your mind when you hear the word (or place name) …?
- Write down four reasons why people might want to visit this place.
- What would a reader of this newspaper want to know about this event/issue? (identifying questions)
- What surprised you most in that video extract?
- What words would you use to describe this picture?
- What changes would you most like to see in this area in the next 10 years?

If we want to value and encourage different types of thinking it is worth using, over a period of time, different types of thought showers which probe reasons, opinions and perceptions as well as factual knowledge.

Stage 2: Pooling ideas

The teacher collects ideas from students in a whole-class activity. All ideas are accepted without comment by the teacher or by other students. Contributions are recorded for all to see, either randomly or sorted into categories by the teacher (either explicit or not yet identified).

Stage 3: Categorising ideas

The ideas are structured in some way. If contributions have been recorded randomly, students can be invited to think of ways in which the ideas could be classified. If contributions have already been classified, students can guess or discuss the basis for classification. It is clearly advantageous if the way the contributions are classified is relevant to the particular investigation. If it seems appropriate, the teacher and/or students make a note of the initial ideas (or save them from an interactive whiteboard).

Stage 4: Generalising

At this stage it is useful to sum up what has been learnt from the activity. Which ideas/perceptions are dominant? Why? What is the origin of the ideas/perceptions?

Stage 5: Making a link with the enquiry questions and enquiry procedure

The teacher provides a link between the thought shower activity and the next activity, showing how students' ideas are contributing to the enquiry.

Example 2: Group spider diagram showing ideas on pollution.

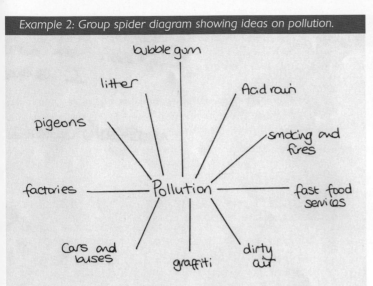

Spider diagrams and bubble diagrams

Spider diagrams have one central word or idea surrounded by related words and ideas. The spider diagram in Example 2 was produced by a small group of year 8 students sharing their ideas about pollution. Their initial ideas are unsorted and their ideas range in scale from bubble gum to acid rain.

Bubble diagrams, in contrast, encourage students to start by identifying broad categories and then to sub-divide these. The bubble diagram in Example 3 was produced by a small group of year 8 students initiating ideas about problems in urban areas.

Spider and bubble diagrams can encourage a more extensive response from each individual than thought showers and they can also provide a written record of prior knowledge which can be used at the end of enquiry work for comparison. They are suitable for topics on which it is likely that students will have considerable existing knowledge.

Example 3: Bubble diagrams help students to sub-divide categories that arise in spider diagrams.

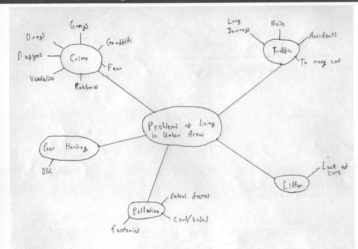

Concept mapping

Some books refer to spider diagrams as concept maps. This book, however, uses the term 'concept map' to refer to a diagram in which the emphasis is on the links rather than the nodes of the diagram (Ghaye and Robinson, 1989; Leat and Chandler, 1996). The main task for students is to identify connections between concepts and to suggest relationships. General instructions for concept mapping are set out in Figure 4. When concept maps are used at the start of enquiry work they can reveal patterns of thinking and sometimes serious misunderstandings. It is useful, therefore, to debrief the concept map activity to probe the reasoning behind the links students have made. Some of the relationships identified can form tentative hypotheses and can be returned to later. The concept map shown in Example 4 was produced in three stages by a small group of year 8 students. First, they identified some key concepts related to problems of living in urban areas. Second, they arranged these concepts on a skeleton outline of a concept map. Third, they tried to identify possible links between the concepts.

Concept maps can also be used at later stages of the enquiry process, either as a way of using data to make notes, or as a way of making sense of information collected from a range of data.

Frameworks to prompt thinking

A variety of frameworks can be used to provide an initial structure to get students started. The framework can take the form of questions. Wray and Lewis (1997), for their research in primary schools, devised a simple grid of three columns headed with the following questions: 'What do I know?' 'What do I want to know?' 'What have I learnt?' Some teachers added a column headed 'Where can I find the

Figure 4: Instructions for concept mapping. Source: Leat and Chandler, 1996.

These instructions for concept mapping were devised by Leat and Chandler. They are designed to be used with a set of concepts, each one written on a card.

1. Sort through the cards, discuss them and put aside any that you don't understand. You will be given an opportunity to check these out with me.

2. Put the cards on the piece of paper and arrange them in a way that makes sense to you. Discuss the possible links between the terms. Those with many links can be kept close together, but do allow space between all cards because more cards may be added later.

3. When you are satisfied, stick them to the sheet.

4. Draw lines between the terms that seem to be connected.

5. Write on the line a short explanation of the link. Use arrows to show which way the link goes. Different links can go in both directions for any pair of terms and there can be more than one link in any direction. There does not have to be a link between all the terms.

6. On the blank card(s) add any missing terms that you think are important and add in the links.

information?' (see 'The EXEL Project and writing frames' above). These KWL and KWFL grids, as they were termed, can be used to draw attention to the research process involved in enquiry.

The framework can take the form of headings or some initial information. The framework in Example 7, Chapter 9 (pages 124-125), provided students with the names of countries in Europe, distinguishing between those within and those outside the European Union. This was used to elicit students' existing knowledge.

Stimulus and response

Another way of eliciting knowledge, ideas, and perceptions at the start of the writing journey is to provide some kind of stimulus, e.g. a short video. Students can be invited either orally or in writing, to:

- suggest descriptive vocabulary
- say what surprised them most
- ask questions, either freely or within a framework, e.g. the five Ws (Nichols with Kinninment, 2001) or based on the Birmingham DEC Compass Rose (see Chapter 3, page 38)
- suggest hypotheses to explain what they have seen
- suggest positive and negative features.

The stimulus and response activity should be designed to motivate students for the written work that is to follow. These initial responses and ideas can be developed in a similar way to that suggested above for thought showers.

Example 4: This concept map, produced in stages by a group of year 8 students, indicates the links between the problems of living in urban areas.

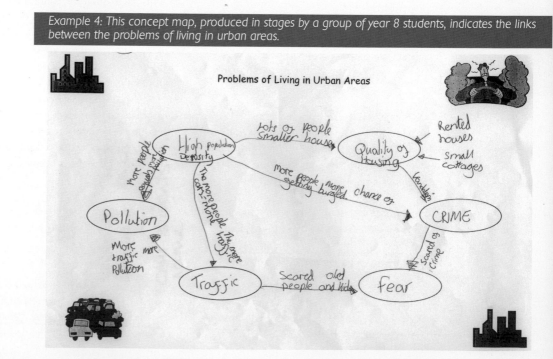

USING DATA: MAKING NOTES

Purposes, difficulties, skills and strategies

When first-hand data is collected during fieldwork, attention is usually given to the types of data needed, and how data are going to be categorised, collected and recorded. Similar attention needs to be given to collecting data from secondary sources. The process of making notes from data serves important purposes in geographical enquiry work. It enables students to:

- collect information (according to some criteria of selection and rejection)
- record information for future use in enquiry work, to be reorganised, e.g. in extended writing or in a role-play to help students make sense of what they are studying.

Teachers can underestimate the difficulties students have in making notes from such data. Sheeran and Barnes noted that students often spent time 'copying considerable chunks of textbook prose' (1991, p. 59) in humanities lessons. They did not always know what was meant by 'writing notes'; different teachers meant different things. They noted that some students 'only mentioned the surface criteria of leaving out words and shortening sentences, without mentioning the idea of homing in on the key ideas in a text'. The ground rules about notes had been interpreted as being about form rather than about making sense. Students also did not know what was relevant, because they were unclear about the conceptual framework that informed the teachers' understandings of relevance. They concluded that students needed support in writing notes.

Students need to develop some specific skills of note-taking, including those of:

- selecting and rejecting information
- sorting information
- abbreviating words
- presenting notes in clear formats (diagrams, or well-organised text).

'Note-taking doesn't make sense unless students have some wider contextual map of the task' (Sheeran and Barnes, 1991, p. 60).

These skills need to be developed in purposeful contexts because the criteria for selection and rejection, the type of sorting needed, and appropriate structures for presenting information all depend on the nature of the enquiry, the questions being asked, and the conceptual framework that will be useful for developing understanding. Support for note-taking can be provided through the use of:

- DARTs activities
- note-taking frameworks
- debriefing the note-taking process.

DARTs

The DARTs activities presented in Chapter 5 (pages 54-58) introduce important note-taking skills including the use of underlining to select key points and the re-ordering of information in simplified tables and diagrams.

Note-taking frameworks

Note-taking frameworks are visual frameworks, structured according to geographical content or concepts, onto which students can map information from data. Students can be introduced to particular note-taking frameworks produced by teachers. If students are to learn to take more control of their enquiry work, then note-taking frameworks can be negotiated with students or developed by students. It would be hoped that eventually students could devise their own frameworks for making notes.

Note-taking frameworks can take the form of tables, pages divided into boxes, Venn diagrams (Example 5b, Chapter 9, page 120), 'T note-taking frame' (Example 5c, Chapter 9, page 120), other diagrams and flow charts. They can include headings or questions used as headings which structure the note-taking. The space available for writing can be limited to encourage selection, simplification and abbreviation. A number of bullet points can be included in the framework to set a target for, yet also limit, the number of key points to be identified. The frame should be structured according to the needs of the enquiry, the questions it is asking and the categories of information required. The different ways of categorising information for analysis and reconstruction DARTs, shown in Figure 1, Chapter 5 (page 55), can be used to structure note-taking frames.

Debriefing note-taking

If students' note-taking skills are to be developed, then time needs to be given to discuss them explicitly. This can be done prior to the note-taking, so that students are provided with a demonstration of what the notes might look like. It can also be done through debriefing the note-taking process. The following procedure is suggested:

1. Students make notes using the same data and the same guidelines.

2. Students, in pairs, compare their notes and identify similarities and differences.

3. The teacher finds out, through whole-class discussion, what kinds of notes students have made, asking questions such as: What did you write in this section? Has anyone written anything different? (List different notes on board or overhead transparency.) Are all these points important? Could any of the points be abbreviated or summarised? Were any parts of the framework difficult to make notes in? Is there any information you found difficult to categorise? How could you make these notes more useful for writing your report/for making your presentation? What have you learnt about note-taking from this discussion?

MAKING SENSE: EXTENDED WRITING

The most common way of presenting the findings of geographical enquiry work is through a piece of extended writing. Extended writing can help students make sense of information they have collected from data, to reorganise it, to make connections between different bits of information, in order to tell some sort of geographical story.

Students need support at this stage of the writing journey, just as they need support in generating ideas and making notes. Ways in which teachers can scaffold the writing process include:

- collaborative whole-class discussion
- support for individuals through discussion
- discussing first drafts
- specific support related to the type of writing required.

Collaborative whole-class activity

In order to support written work teachers need to do more than explain to students what is required. The teacher and students need to discuss and think together about:

- the purposes of writing and how it relates to the original enquiry questions. What geographical 'story' is the extended writing trying to tell?
- choices available to students in terms of audience, form, etc.
- what is expected. What are the ground rules? What would be considered a successful piece of writing? How is it going to be assessed?
- examples of similar types of writing, from the world outside the classroom, from geography texts or from students, including good and weak examples
- possible ways of structuring the writing
- phrases and vocabulary.

Support for individuals through discussion

Although the amount of support that can be given to individuals in a classroom is limited, some support can be given during the writing process. Students are likely to need help with both the what and the how of writing, with both the meaning and the form, so teachers should make a point of discussing both. Dialogues between teacher and individual students can draw attention to:

- positive features of what is written and its geographical meaning
- positive features of the form of the writing, e.g. correct spelling, good use of paragraphs
- negative features of what is written geographically, e.g. errors, misunderstandings, lack of clarity
- negative features of the form of the writing, e.g. spelling and grammatical errors
- the student's thinking.

It is all too easy to focus on the negative features of writing, even if the dialogue is constructive. As writing is a difficult and generally unpopular classroom activity, it is worth ensuring that students receive as much positive feedback as possible. Not all dialogues, however, need to be judgemental. The teacher can simply show an interest in what the student is writing and try to understand how the student is thinking. It is also worth ensuring that much of the dialogue focuses on geographical understanding.

The individual dialogues about writing between teacher and student carry messages to the students about what is important. The ways in which dialogues between teacher and student support learning, and sometimes hinder learning, would be a good focus for action research.

First drafts

The process of drafting and then modifying a piece of writing to make the meaning clearer and to correct errors is valuable and should be used occasionally, particularly if the written work is for 'publication' in a display or a booklet for others to read. Discussing first drafts in pairs or groups can make students more aware of how to make writing make sense and of the expected conventions of correct spelling and grammar.

Specific support related to the type of writing required

Extended writing in geography can take a variety of different forms including: a report, a descriptive account, or other forms of impersonal geographical writing. It can also include other forms of writing which are important in the 'real world' for communicating geographical information, e.g. newspaper articles, CEEFAX, brochures, advertisements, posters, letters, postcards, stories. In order to succeed academically, students are expected to be able to write in a wide variety of styles or forms. In GCSE Geography, for example, students can be expected to write accounts of fieldwork, to compare places, to describe and explain processes, to express opinions and justify choices, and sometimes to write letters to imagined newspaper editors. Students need to develop their skills in writing in different forms throughout their study of geography in school.

The National Literacy Strategy's approach to developing these skills has been to introduce students to the concept of text types, each with their particular linguistic characteristics. This is intended to help students to select and organise information in more effective ways. Students at key stage 3 will already have been introduced to the following text types:

- instructions
- recount
- explanation
- information
- persuasion
- discursive writing
- analysis
- evaluation.

The National Literacy Strategy expects students to learn to use these different text types, in purposeful contexts. The Key Stage 3 Strategy encourages the use of writing frames for different text types as a way of scaffolding students' writing and suggests a range of starter sentences and connectives associated with each text type (DfEE, 2001a).

The extent to which knowledge of different text types and the use of writing frames is an appropriate way to improve extended writing is a contested issue. Some researchers are sceptical about whether teaching about the surface forms of different text types is the way to improve writing (Sheeran and Barnes, 1991; Webster, 1996; D'Arcy, 2000). They would put more emphasis on the context in which writing takes place, on the meaning rather than the surface features of text and on students being able to challenge as well as conform to expectations about writing.

Writing frames were originally devised for primary school children, particularly for those with literacy problems (Lewis and Wray, 1995). They were intended, like scaffolding, as temporary support where needed. They were intended to be used in purposeful contexts following whole-class discussion. They need to be used with care by secondary school teachers. Writing frames can control and limit what students write as well as enable them to structure their writing.

Greig (2000) showed how writing frames could be used to provide differentiated support for learning. His students had studied floods in Bangladesh through a mystery activity. During the mystery activity they complete a simple note-taking frame with two headings:

- first thoughts and findings
- further thoughts and findings.

Following the mystery activity, students were asked to produce a written account answering the question: 'Why did the flood of 1998 have such a big effect in Bangladesh?'. He provided two differentiated writing frames:

- a writing frame offering more support, with both sentence starters and a word bank,
- a writing frame offering less support with suggestions for each paragraph and a word bank.

The emphasis in Greig's work was on developing geographical understanding. Figure 5 provides some guidelines for using text types and writing frames.

REFLECTING ON LEARNING

The use of learning journals

The culture of schooling generally encourages students to conceal ignorance and misunderstandings. This is particularly true of written work, where assessment criteria are related to evidence of positive features of learning such as understanding, coherence and accuracy. If, however, we really want to help students to learn, it is more helpful to know what they do not understand, what they found difficult and what they are still confused about. As part of the enquiry work students can be encouraged to write learning journals (sometimes referred to as diaries or learning logs) in which they can communicate thoughts about their learning to the teacher. Example 5 shows a year 7's completed learning log. The student was investigating 'a fair view of England' (details are shown in Example 2, Chapter 14, pages 184-186). This was the first time that these year 7 students had written learning logs so support was

---

Figure 5: Guidelines for using text types and writing frames in geographical enquiry work.

For each piece of extended writing related to a particular enquiry and enquiry question, the teacher:

- considers the geographical purposes of the enquiry and the types of questions it is exploring

- considers what type or types of writing would be appropriate to explore these questions, e.g. one text type, a combination of text types or other forms of writing, e.g. newspapers, letters or brochures

- considers what choices about form of writing should be available for students

- emphasises, in the discussion preceding extended writing, what needs to be communicated, to whom and why, stressing the meaning of the writing and the geographical story it is going to tell

- introduces students to examples of the kinds of writing expected for this enquiry and the ways in which they convey meaning

- offers differentiated support in the form of sentence starters and word banks

- offers differentiated support through discussion with individuals

- reflects, after the lesson, on ways in which support for the written activity has improved students' geographical understanding and ways in which support might have limited it (a possible focus for action research).

Example 5: An extract from one student's learning log.

Learning Log for 7HX!

During my geography lessons with Miss Bird and Miss Wade I have learnt about... *England having many different scenic locations and to see all that England offers it would take a life time. Also it has improved my computer skills.*

The things I liked about these lessons are... *Using the computers, copying from the internet and planning an enjoyable holiday for someone else.*

I now understand... *it is important to show the good side of everything but an equal amount of bad.*

The things I did not like about these lessons are... *adverts popping up on the internet.*

I do not understand...

I would also like to write about geography... *in countrys that are at war or that are near the equater, for they are very different to our own. And I would enjoy planning another holiday that must not use places that we used in this lesson.*

provided in the form of sentence starters. These encouraged evaluative comments on what the students liked and disliked and reflective comments on what was understood and what not understood. Interestingly, no students using this framework completed the sentence starting, 'I do not understand ...'. Perhaps all students understood everything! Or perhaps it takes time to encourage students to reveal or indeed to articulate what they do not understand.

ASSESSING AND USING WRITTEN WORK

Black and Wiliam's research showed that improving learning through assessment depended on five factors:

- *'the provision of effective feedback to pupils*
- *the active involvement of pupils in their own learning*
- *adjusting teaching to take account of the results of assessment*
- *recognition of the profound influence assessment has on the motivation and self-esteem of pupils, both of which are crucial influences on learning*
- *the need for pupils to be able to assess themselves and understand how to improve'*

(1999, pp. 4-5).

If we apply these findings to the assessment of written work completed as part of the process of enquiry then we need to involve students in the assessment process right from the start. They need to be involved in discussions on what makes a good enquiry, a good report, and a good piece of writing before they start writing, not after the work has been marked and completed. They need to be aware of the criteria for assessment and they need to be able to apply these to their own work.

When written work is marked, the 'marks' on their work need to be constructive. If the purpose of enquiry is to promote understanding at the same time as developing enquiry skills, then attention needs to be given to the geography of what is written and how effectively enquiry skills have been used, as well as to the linguistic features of spelling, punctuation and grammar.

Black and Wiliam (1999) devised a quick way for students to communicate their own self-assessment to the teachers. Students marked their work with 'traffic light' circles – green for when they understood a topic well, orange to indicate incomplete understanding and red to indicate confusion. Teachers involved in the project found this really helpful in that it alerted them to when students needed more support. The 'traffic lights' strategy has potential for use in geography.

Using written work

Written work is given greater value if some use is made of it after it has been 'marked', and if it reaches a wider audience. There are several ways in which this can be achieved:

- including written work in displays, either on sheets of paper or in small booklets
- producing a class folder of work, accessible to all students and possibly at a parents' evening, on a particular theme
- writing for real audiences, for people other than the teacher, who will receive the work and possibly respond to it.

SUMMARY

Writing can enable students to make sense of what they are investigating throughout the enquiry process. Geographical enquiry can provide a purposeful context for each stage of the writing process; it can provide a purposeful context for initiating ideas, for making notes, and for extended writing. Many students find writing difficult so it needs to be supported at every stage of the writing process with attention being given to both the geographical meaning and the form of the writing.

INTRODUCTION

This chapter sets out to examine why classroom talk is thought to be so important for learning. It outlines some of the research into classroom talk before presenting and discussing strategies to develop talk as an integral part of geographical enquiry. Although the focus of this chapter is on speaking and listening, most of the classroom activities discussed will also involve reading and writing.

RESEARCH

'One of the most important ways of working on understanding is through talk, either in formal education or as part of the learning in everyday life. New ways of talking about things lead to new ways of seeing them'

(Barnes and Todd, 1995, p. 4).

The importance of talk in the learning process has long been recognised. Psychologists distinguish between two functions of talk as a means of:

- communicating with each other, sharing and developing the cultures within which we live
- making sense of the world for ourselves, for organising our own thoughts as we speak and for reflecting on what we have done.

Vygotsky (1962) stressed the role of language, but particularly the role of talk, in making sense of the world, in learning. He thought that it was through dialogue with supportive adults that children progressed to higher levels of thinking (see 'Reflections on practice', page 15).

The Bullock Report (DES, 1975) emphasised the importance of students' talk in promoting learning. Its publication acted as a catalyst for a substantial amount of research into classroom talk across the curriculum, both in whole-class and small-group situations.

Research into whole-class talk has consistently shown that it is dominated by teachers. Teachers ask most of the questions, the majority of which are low level ones demanding the recall of information but little thinking. Barnes (1976) showed that in typical question and answer sessions it was teachers not students who made sense of information; it was teachers who controlled the construction of knowledge. Mercer (1995), building on Barnes' work, has investigated how classroom talk could be used to create joint understanding or 'common knowledge'. Mercer's recent work (2000) identified characteristics of teachers who use talk effectively in the classroom (see panel below). His research showed that teachers who were effective encouraged students to reason, to make explicit their own thought processes. He emphasised the importance of exchanging ideas in the classroom to build up 'common knowledge'.

Characteristics of effective teachers.

Neil Mercer (2000) was involved in work carried out by Sylvia Rojas-Drummond and her colleagues at the University of Mexico researching education in primary schools. They compared teachers who achieved good results in reading comprehension and problem solving in mathematics with teachers working with similar classes with less good results. After lengthy analysis and interpretation of video recordings, they concluded that the more effective teachers shared these characteristics:

1. *'They used question and answer sequences not just to test knowledge, but also to guide the development of understanding. These teachers often used questions to discover the initial levels of students' understanding and adjust their teaching accordingly, and used "why" questions to encourage students to reason and reflect on what they were doing.*

2. *They taught not just "subject content", but also procedures of solving problems and making sense of experience. This included teachers demonstrating to children the use of problem-solving strategies, explaining to children the meaning and purpose of classroom activities, and using their interactions with children as opportunities for encouraging children to make explicit their own thought processes.*

3. *They treated learning as a social communicative process. This was represented by teachers doing such things as organizing interchanges of ideas and mutual support amongst students, encouraging students to take a more active, vocal role in classroom events, explicitly relating current activity to past experience and using students' contributions as a resource of building the "common knowledge" of the class.'*

(Mercer, 2000, pp. 159-60).

Barnes and Todd (1995) found that students working collaboratively could use exploratory talk to enhance their conceptual understanding. They showed that small-group work gave students opportunities to present ideas tentatively, test them against the ideas of others and reshape their understanding. Reid *et al.* (1989) suggested a sequence of activities that would encourage the development of understanding through talk (see panel below). The sequence includes the use of two types of talk: exploratory talk, through which students make sense for themselves, and presentational talk, in which students have to communicate to an audience.

Between 1987 and 1991 a major national project, the National Oracy Project (NOP), developed strategies for both exploratory talk and presentational talk through work in 35 local education authorities (Norman, 1992). The project, aimed at developing students' oracy skills in all subjects of the curriculum, stressed the importance of oracy being developed during worthwhile activities rather than as a separate skill. Groups of teachers worked together, developing and sharing action research projects. The group work in Staffordshire, for example, developed a range of groupwork activities across the primary and secondary curriculum, including geography. They found that students working in small groups were able to discuss enthusiastically and that teachers learnt a lot from listening more and through their action research. One teacher commented: 'I have witnessed a quality of thought and process that I would normally be unaware had taken place' (Carter, 1991, p. 68). The potential impact of NOP was impeded by the introduction of the national curriculum.

All the research suggests that speaking and listening have a crucial role in helping students to develop their understanding. The strategies presented and discussed below are not new; they have been informed by a large amount of research into classroom talk generally, some of it carried out in geography classrooms. I have categorised the strategies into three groups:

- strategies for using speaking and listening in whole-class situations
- strategies used in small-group work
- enquiries in which speaking and listening are dominant.

Stages in using talk to promote understanding.

Reid *et al.* (1989) researched small-group work in classrooms in Australia and identified a sequence of activities that would promote learning: engagement, exploration, transformation, presentation and reflection. Barnes and Todd (1995) thought that these five stages could be used to inform planning of sequences of lessons as well as small-group work taking place in one lesson. They interpreted what the stages meant, as follows:

'*Stage 1: Engagement – refers to the provision of experience or information and includes the arousing of the students' interest.*

Stage 2: Exploration – is likely to be a kind of thinking aloud, an initial searching through the new experiences or information in relation to what the students already know and understand.

Stage 3: Transformation – refers to the stage at which the teacher asks students to work upon the new material, for example by clarifying, re-ordering, elaborating and perhaps applying it to various purposes. This seems likely to be the stage at which students do most to make the new material or way of thinking their own.

Stage 4: Presentation – implies that an essential stage in group work is the reordering for an audience of new material produced during the transformation stage.

Stage 5: Reflection – refers to an opportunity for the student to reflect explicitly both on the content of the sequence of unit or learning, and on the learning process involved in the group discussion they had experienced'

(Barnes and Todd, 1995, p. 85).

These five stages incorporate the essential ideas of a constructivist theory of learning, in which students learn by being actively involved in making sense of things for themselves in relation to what they already know.

SPEAKING AND LISTENING IN WHOLE-CLASS SITUATIONS

Establishing the ground rules for speaking and listening

If students are to use discussion as part of geographical enquiry, it is important that a classroom environment is created in which students are able to develop both the social and the cognitive skills that are needed to learn from discussion. Socially, it is important to create an atmosphere in which all students are able to express their thoughts, views and feelings without anxiety and in which students listen to each other with respect. Cognitively, it is important that students learn to give reasons for what they say and that they are able to probe what others say.

This kind of classroom environment is not going to happen accidentally; it needs to be created. Whenever classroom talk is used as a means of enquiry, whether it is for whole-class discussion, a formal role-play, a visiting speaker, or small-group work, attention needs to be given initially to the ground rules for discussion.

Mercer (2000) found that the success of 'talk lessons' in primary schools depended on teachers creating 'communities of enquiry' in their classrooms. He found that in the early stages of establishing the talk lessons, it was important to discuss and agree explicit ground rules with the children. The sets of ground rules agreed by two classes are shown in Figure 1. Example 1 describes how a secondary school in Sheffield established ground rules to create successful 'communities of enquiry' in PSHE lessons. Two clear messages emerge from the research and from this Example:

- it is worth taking time initially to establish ground rules
- the involvement of students in producing the ground rules is a key to success.

If we are to make the most of using talk as part of geographical enquiry, then it is worth geography teachers taking time at the beginning of key stage 3 to negotiate the ground rules of classroom talk with their students.

Figure 1: Ground rules for classroom talk. Source: Mercer, 2000, p. 162.

The following sets of ground rules for classroom talk were agreed in two Mexican primary schools in a 'Talk lesson' project developed by Neil Mercer in collaboration with Sylvia Rojas-Drummond.

Our ground rules for talk

We have agreed to:

    Share ideas

    Give reasons

    Question ideas

    Consider

    Agree

    Involve everybody

    Ensure that everybody accepts responsibility

Our talking rules

We share our ideas and listen to each other

We talk one at a time

We respect each other's opinions

We give reasons to explain our ideas

If we disagree we ask 'why?'

We try to agree in the end

Stance

'Stance' is about the way teachers talk throughout lessons. Bruner described in *Actual Minds, Possible Worlds* (1986) how the 'stance' of one of his teachers towards knowledge inspired him:

'I recall a teacher, her name was Miss Orcutt, who made the statement in class, "It is a very puzzling thing not that water turns to ice at 32 degrees Fahrenheit, but that it should change from a liquid into a solid." [Bruner comments] 'She was inviting me to extend my world of wonder to encompass hers. She was not just informing me. She was, rather, negotiating the world of wonder and possibility. Molecules, solids, liquids, movements were not facts: they were to be used in pondering and imagining. Miss Orcutt was the rarity. She was a human event, not a transmission device' (Bruner, 1986, p. 126).

Bruner states that we have a choice of stance: we can 'close down the process of wondering by flat declaration of fixed factuality' or we can 'open wide a topic to speculation and negotiation'.

Example 1: Ground rules for lessons. Source: King Edward VII School, Sheffield.

(a) Establishing ground rules

'The school's provision for pupils' personal development is a major strength. Through its unique ethos, it places great emphasis on pupils' personal responsibility in the learning process' (Ofsted, 2002, p. 18).

Personal, social and health education (PSHE) at King Edward VII School in Sheffield is highly successful, praised by students and visitors alike. PSHE occurs throughout the school through the pastoral system, in all lessons and through its ethos. It also exists as a specific subject on the timetable, taught by three specialist teachers to mixed ability classes for students aged 11-16. What is distinctive about these timetabled lessons is that the students learn almost entirely through talking with each other.

The teachers in charge of PSHE attach great importance to building good relationships between students and to negotiating ground rules for discussion. Year 7 students come from over 30 feeder schools, so both teachers and students need to get to know each other at the start of the year. The PSHE department therefore uses the first four lessons of the course to begin to create a good classroom environment for learning through speaking and listening.

Lessons 1 and 2 are used for 'getting to know each other' activities. Lesson 3 is most uncharacteristic of PSHE lessons; students work individually in silence and they write. They complete the sheet of sentence starters on a sheet headed, 'I'm comfortable!' (see Example 1b).

The teacher collects the students' responses and, before lesson 4, compiles a list of ground rules for that particular class based on what the students have written.

In lesson 4, students discuss what each 'rule' on the list means. If the teacher feels that something essential has been missed out, e.g. the need for confidentiality in relation to what is said in the lesson, then she or he suggests it to the class for consideration. Their responses, plus any amendments or additions, become the agreed ground rules for that particular class and are referred to throughout the year. Example 1c shows the ground rules produced by one year 7 class.

At the beginning of years 8 and 9, when students are taught PSHE in the same groups, the ground rules established the previous year are looked at in the first PSHE lesson of the year, and renegotiated. In year 10, new PSHE groups are formed and each class completes the sentence starter activity again to establish the ground rules for years 10 and 11.

Ground rules on their own of course are not enough. The success of King Edward VII PSHE lessons is also dependent on:

- the enthusiasm and total commitment of the specialist teachers
- the layout of the room: there are no desks, and chairs are arranged in a circle
- lessons taking place in attractive classrooms, with displays, plants and resources available
- use of small-group work before whole-class discussion
- use of types of activities likely to promote discussion in small groups, e.g. diamond ranking
- use of groupwork strategies that heighten awareness of role, e.g. listening triads
- frequent use of praise
- visiting speakers being asked to answer questions from students instead of making presentations
- the role of the teacher as facilitator of discussion.

Students' comments on the course are favourable and perceptive, for example:

- 'I've learnt to listen to other people.' (Y7)
- 'I feel that the work we have done in our Y8 PSHE lessons has really taught me a lot. It's my favourite lesson of the week. It goes into details of life and reality. If we didn't have this lesson there would be so much I wouldn't understand.' (Y8)
- 'In PSHE we learn how to communicate with people and work in groups.' (Y9)
- 'I think PSHE helped me prepare for work experience.' (Y10)
- 'In PSHE we learn how to be more understanding and more confident about ourselves.' (Y7)
- 'A lot of things I already knew were expanded and some things, which I thought I knew, were corrected as I had the wrong information.' (Y10)
- 'I feel that the work has given me a greater understanding of how the system of the law works and what my rights/responsibilities are.' (Y10)
- 'The fact that people had the chance to say their points of view and opinions even if they didn't agree with mine.' (Y10)
- 'Respect for yourself and others ... that's what you learn in PSHE.' (Y8).

These comments on a course taught almost entirely through talk show that speaking and listening can be a powerful means of learning; students have learnt invaluable social skills at the same time as increasing their knowledge and understanding.

Note: King Edward VII School, Sheffield, is an inner-city comprehensive school with over 1600 students on roll and a sixth form of almost 500 students. The intake of students into year 7 comprises a wide social mix, coming from more than 30 different primary schools. Over 25% of students are from ethnic minority backgrounds.

Example 1: Ground rules for lessons. Source: King Edward VII School, Sheffield ... continued

(b) Negotiating the ground rules of discussion

Students complete the following sentences:

I'm comfortable!

- I feel best with others when they …
- I feel uncomfortable with others when they …
- It is easy to share my feelings with others when they …
- It is hard to share my feelings with others when they …
- In our class I hope that …
- In our class I am most concerned about …
- What I want to get out of these lessons is …
- I feel warmest towards others when they …
- What I can give to this group is …
- I would like to discuss …

(c) Ground rules produced by one year 7 class for PSHE

We do want people to …

talk with me
include others
listen
be warm
be nice and kind
help others
show respect
be considerate
be helpful
be trustworthy
care and be friendly
get along
co-operate
become confident
gain social skills
encourage
show interest
show understanding
share views/feelings
be open
be happy
argue well
have fun
have a joke
get to know others

We don't want people to …

make fun of others
be left out
bully
giggle
feel embarrassed
be horrible/pick on others
distract others
exclude people
feel ignored
be secretive
laugh at others
tease
be mean
feel shy
be put under the spotlight
whisper
be racist
bug others
use put-downs
talk behind people's backs
mess about
spoil the lesson
stop others learning
lie

In enquiry work it is important to 'open a topic to speculation' rather than to declare 'fixed factuality'. It is possible to open up 'the process of wondering' in relation to each of the essential aspects of enquiry work.

- *Creating a need to know:* Do we tell students beforehand what they are going to learn? Or do we sow seeds of curiosity in students' minds, and wonder what sense they will make of the data? If we do the former by providing detailed learning objectives, what role is there for students in the construction of meaning? What if their understandings are different?

- *Using data:* Do we present data to be accepted as the truth? Or do we present data as selected evidence to be interpreted and evaluated? If we do the former, how can students begin to understand how the story of geography is created from the interpretation and linking of information rather than from its acceptance?

- *Making sense* and *reflecting on learning:* Do we want students to think that what they have learnt is fixed and certain? Or do we want them to recognise that knowledge is provisional and open to critical debate?

If we talk in factual, certain ways then we convey one set of messages to students. If we talk in puzzled, uncertain, provisional ways then we convey another set of messages. Students pick up these messages, part of the hidden curriculum of the classroom, more easily than they understand the substance of what they are learning. They learn whether they are expected to accept or whether they are expected to question and this will affect the way they approach geographical enquiry.

To promote an enquiry approach we need to open topics to speculation and negotiation, not just at the outset of enquiry work, but throughout. We need to stimulate curiosity. We need to promote a questioning attitude towards data, using data as a selection of evidence, a representation of reality, open to question and debate. We need to make students aware that it is their thinking that enables them to make sense of things, and that they need to develop their understanding in discussion with others. We need to be geographical equivalents of Miss Orcutt.

SPECULATIVE TALK

Stance is about the kinds of talk the teacher engages in and this includes speculative talk. Students can also engage in speculative talk. It is valuable in that it encourages higher order thinking skills and stimulates a need to know. Speculative talk is provoked by questions such as:

- Where do you think this photograph was taken?
- Which countries in the world do you think would …?
- What might this place be like?
- How would we find that out?
- How might we explain that?

Speculative talk is probably most easily encouraged by the use of a picture, video extract, and activity as a stimulus. Example 2 shows how students used speculative talk to answer the question, 'Where in the world?' In Example 3, students speculated on what a particular place in Antarctica would be like. There are further examples of ways of stimulating speculative talk in the 'Intelligent guesswork' activities in Chapter 8 (pages 101-102).

Example 2: Speculative thinking: Where in the world?

This example was used as a starter activity in a unit of work on Japan. The stimulus was provided by 20 pictures of Japan, collected from the internet, which included rural and urban scenes and traditional and modern views of Japan. The activity consisted of four stages:

Stage 1: Introduction to the task

The teacher informs the class that they are going to see 20 pictures. (These can be shown as slides, as transparencies, on a computer screen or projected onto a screen from a computer.)

Task: Decide in which country each picture was taken

Write down the name of the country and next to it what 'clues' suggested this country to you.

Stage 2: Stimulus

The students look closely at the pictures and make notes.

Stage 3: Speculative talk

The teacher goes through each photograph in turn, asking for students to suggest names of countries and to give their reasons. The teacher encourages contradictory contributions and reasoning.

Stage 4: Debriefing discussion

The teacher informs the class that all the photographs were taken in Japan.

Debriefing prompts: What have they learnt about Japan from these pictures? How would they describe it? Which pictures surprised them and why? Which pictures did not surprise them and why? Where have they got their impressions of Japan from? Do they think that these pictures give a reliable impression of Japan? How would they find out? What kinds of information in the pictures were most useful as clues?

Variations on this activity

- This activity can be used for any country in which there are big contrasts.
- Hopkirk (1998) describes the use of a similar strategy with a set of slides all taken in India. He asks students to identify which slides have been taken in a 'more economically developed country' and which slides have been taken in a 'less economically developed country'.
- It can be used as part of the study of thematic topics, e.g. population density.

Note: The images section of www.google.co.uk is an excellent source for places and geographical themes.

Example 3: Speculative thinking: What might it be like?

This example has been used as an introductory activity in the first lesson of a unit of work on Antarctica. This lesson is an example of the use of speculative talk in a descriptive enquiry, aimed at answering the question, 'What is Antarctica like?'

Starter

- The teacher informs the class that they are going to Antarctica with Michael Palin to find out what Antarctica is like. They are going to travel from Chile to Patriot Hills (write Patriot Hills on the board).

- For homework they are going to write a diary entry about the journey to Patriot Hills. (There will be time to discuss this after the video.)

- The students watch a ten minute extract from the BBC video, *Pole to Pole* (1992), starting when passengers board the airplane in Chile.

- Important: The video is stopped just before the airplane lands. (This extract is good, because the anxiety and expectation of the passengers creates a great sense of anticipation.)

Activity: Speculative discussion

- Students are invited to speculate. The plane is due to land at Patriot Hills. 'What will Patriot Hills be like?'(In pairs at first, then feeding their ideas to the whole class.)

- All ideas are listed on the board, possibly arranged in categories. (Students are likely to mention landscape, buildings, weather and people.)

- The teacher encourages responses in different categories of information, e.g. landscape, buildings, with speculative prompts: I wonder what the weather will be like? What do you think? Does everyone think this? How cold do you think it might be? Why do you think that? (The teacher maintains a speculative stance, wondering what the weather will be like.)

- The teacher invites comments on whether any of these ideas are more or less likely and why students think that.

Activity: Close observation of video

The teacher asks the class if they want to see what Patriot Hills is really like (they always will).

Students are asked to watch the next 10 minutes of the video, concentrating both on what they see and what they hear.

Plenary debriefing the observation of the video

From the evidence of the video: What did they find out about the landscape, the buildings, weather, people? Did anything surprise them? Would they like to go there? Why or why not? How much can they tell about what Antarctica is like from this video extract? In what ways might it be misleading? How could they find out more about what Antarctica is like? What evidence would they need?

Activity: Preparation for writing diary entry

Students work in pairs discussing what kinds of things they would write about this journey in their diary.

Students share these ideas with the whole class.

The teacher could read a short extract from a different section of Michael Palin's book *Pole to Pole* (1992). Does this suggest any other types of things to include?

What criteria should be used for marking the diary. How should marks out of 10 be allocated?

Other possibilities

It is possible to use this sequence of activities with other videos. The role of speculative talk is very important in this enquiry in creating curiosity and negotiating which aspects of the video are going to be the focus of attention.

Resources

BBC (1992) *Pole to Pole With Michael Palin* (video). London: BBC Enterprises Ltd.

Palin, M. (1992) *Pole to Pole*. London: BBC Publications.

Focused listening

Occasionally the data or evidence for enquiry is presented orally. This could include:

- a reading by the teacher from newspapers, brochures, travel literature or fiction
- the commentary of a video recording
- a recording of a radio weather forecast
- a presentation by a visiting speaker or by other students
- a tape of an interview.

If we want students to collect information from this type of data, so that they can use it later for making sense of the topic for themselves, then there needs to be focused listening. This is helped if students have:

- already had their curiosity aroused
- some idea of what kinds of information they are listening for
- a note-taking frame providing guidelines on categories of information
- an opportunity to listen more than once
- an opportunity to ask questions to clarify and supplement their notes.

Presentations

Presentations to the rest of the class, or to a wider audience, are a way of communicating the findings of enquiry work. They can provide a focus for group work, encouraging students to reorganise and reshape the information they have in such a way that a larger audience, the whole class, can understand what sense they have made of it. In contrast to the exploratory talk of small groups, with its hesitations, presentational talk is more explicit. If all students are to have the opportunity to participate, then presentational talk needs to be concise. This encourages students to identify the key ideas. Presentations from small-group work will vary according to the activities but might include some of the following:

- Stating the focus of the work (where each group has a different focus)
- Explaining any choices, rankings or decisions made
- Presenting information in an interesting way
- Commenting on the information and what they have learnt from it
- Evaluating the data.

There are many possibilities from the world outside the classroom to provide imagined audiences for presentations, for example:

- Weather forecasts, restricted to 90 seconds, to replicate the 'real world'
- Items for a news report on, for example, an earthquake, a proposed change, locational options for a new industry or for a world sports event – limited to 2 or 3 minutes
- Item for a holiday programme – limited to 2 or 3 minutes
- Presentation for a role-play public meeting (see Chapter 11, pages 145-149.)

Ideally, presentations from different groups should be on different aspects of what is being investigated, to avoid what could become tedious repetition.

Another way of avoiding the repetitious reporting of small-group work is for each group to send an 'envoy' to another group to report on the group's activity. This strategy still demands presentational talk from the envoy, but in a less daunting situation than whole-class presentations.

DEBRIEFING

Debriefing is a term widely used in training sessions in the world of work as well as in education, to refer to the reflective discussion taking place after an activity or an experience. In geographical education, ideas about debriefing have been developed mostly in relation to development education activities, e.g. *The Trading Game* published by Christian Aid, and in relation to 'thinking skills' (Leat, 1998; Leat and Nichols, 1999).

Figure 2: Characteristics of talk in question and answer and debriefing activities.

Typical question and answer activities	Debriefing activities
Most questions are closed – demanding 'right' answers	Most questions are open – inviting a range of answers
The emphasis is on lower order questions, demanding recall of facts or simple ideas	The emphasis is on higher order questions, demanding linking of ideas, reasoning, speculating
The teacher is interested in finding out if the students have understood what is in the teacher's mind	The teacher is interested in what is in the students' minds
The majority of the talk is done by the teacher	The majority of the talk is done by students
Typical student contributions are very short, often one word	Students talk at greater length, developing ideas
The emphasis is on what has been learned	Attention is given to both what has been learned and how it has been learned
The teacher asks questions	The teacher both asks questions and makes statements
The teacher judges every answer	The teacher uses answers to develop the discussion

Although the kind of talk that takes place in debriefing sessions is similar to that used in the PSHE lessons in Example 1, it is very different from typical question and answer sessions in geography lessons (Figure 2). The skills of debriefing are not really difficult in themselves, but go against deeply ingrained patterns of classroom practice and against what students expect. It helps therefore to be explicit to students about the purposes of debriefing, to develop a repertoire of questions and comments that encourage student talk and, until debriefing skills have become as intuitive as questioning skills, to plan for the debriefing session with particular questions and comments and to evaluate its success.

Leat and Kinninment (2000) investigated the value of debriefing for 'thinking skills' activities and identified 'powerful features' of debriefing that promote learning (Figure 3). Students interviewed by Bright and Leat (2000) following the use of a mystery activity were all positive about debriefing:

Figure 3: Potentially powerful features of debriefing. Source: Leat and Kinninment, 2000.

Leat and Kinninment (2000) found that the following characteristics of debriefing made it a powerful tool for promoting learning:

- 'maintaining a high proportion of open questions which require the articulation of reasoning by pupils

- encouraging pupils to extend and justify their answers, if necessary giving thinking time

- encouraging pupils to evaluate each other's contributions to discussion

- providing evaluative feedback to pupils, not necessarily in the form of "that was good/not so good", but in terms of criteria that pupils can apply

- using analogies, stories and everyday contexts to help pupils to understand the wider significance of their learning and encourage them to transfer it

- drawing attention to the cognitive and social skills that the pupils have used and encouraging distillation of good practice in relation to these skills

- relating thinking and learning described by pupils to the important concepts or reasoning pattern in subjects

- drawing on what one has heard or seen during small-group work preceding the debriefing.'

- 'The debrief helped me understand it best.'
- 'It removed muddle.'
- 'I didn't really understand the situation at first – after the discussion I understood it.'
- 'Our ideas weren't going anywhere at the beginning, we didn't know where to start – discussion finally sorted them.'

Debriefing has an important role not only in debriefing activities related to thinking skills and games such as *The Trading Game* (Christian Aid, 2001) but also for all types of geographical enquiry work. One of the characteristics of debriefing is its responsiveness to what students are saying and thinking, so it is impossible to plan for it in detail. General questions that encourage a questioning, critical stance towards enquiry could include:

- What were we trying to find out?
- How did we set about doing this? Was this a good way of finding out? Why? What other ways could we have used?

- How useful/reliable was the data you used? Why? What other data might have been useful? Why?
- Might other data have given different information?
- What have you learnt from all this information? Are there things you are still unsure about? Have you changed your views in any ways? Why?
- What else would you still like to find out?

Research suggests that the kind of talk that takes place in debriefing activities should not be limited to discussion at the end of lessons or even to whole-class discussion. Webster *et al.* (1996) concluded from the large-scale research project they were involved in that this kind of talk is the 'difference that makes the difference' to what is learnt. Mercer argued that this kind of talk enables students to be 'apprentices in thinking, under the expert guidance of their teacher' (2000, p. 161). It should take place between pairs, small and large groups, as well as with the teacher.

The kind of classroom talk in which teachers really listen to what students say, in which teachers are interested in getting inside students' minds in an attempt to understand what and how students are thinking, should be at the heart of all teaching and learning. This is the kind of talk that will take students to higher levels of thinking. It is essential at all stages of enquiry and not just in a summary debriefing session.

SPEAKING AND LISTENING IN SMALL GROUPS

Small-group work provides opportunities for students to use exploratory talk rather than presentational talk to help them learn. It has been defined as follows:

'Exploratory talk is that in which partners engage critically but constructively with each other's ideas. Relevant information is offered for joint consideration. Proposals may be challenged and counter challenged, but if so reasons are given and alternatives are offered. Agreement is sought as a basis for joint progress. Knowledge is made publicly accountable and reasoning is visible in the talk' (Mercer, 2000, p. 98).

Exploratory talk in small groups is often hesitant, tentative and incomplete as students test out ideas with others, rethink what they already know and try to reach some common understandings. Small-group work is one of the best ways of enabling students to do this, but it needs to be well managed, it needs appropriate activities and can benefit from the use of particular strategies, which will be discussed below. It also needs to be part of a purposeful sequence of activities, as suggested in 'Characteristics of effective teachers' (pages 80-81).

Managing small groups

Research does not suggest that there is one correct way of managing small-group work in terms of size of groups, composition of groups or time. Teachers have used groups of 2, 3, 4, 5, and more students successfully. Barnes and Todd (1995) recommended a group size of 3 or 4, but they noted advantages of both small and larger group sizes. The smaller the group the greater is the potential for involvement of each individual. The larger the group the greater is the potential for diversity of views and pooling of ideas. If group sizes larger than 4 are used, it is advisable to allocate particular roles to students within the group and to structure the activity to ensure that all of them take part in discussion.

Students often find it more comfortable working in friendship groups, but such groups 'can seek consensus at the expense of a rigorous examination of the topic in hand' (Barnes and Todd, 1995, p. 93). Working with different people can enable students to encounter more new ideas and opinions and encourage them to explain themselves more clearly. Teachers can have particular reasons for selecting students for groups, related to differentiated activities, the need for different gender or cultural perspectives, etc. Ultimately it comes down to professional judgement by teachers who know individuals in a class well. Group work can vary from short two-minute discussions in pairs to collaborative work spanning several lessons. It depends on the reason for using group work and the nature of the activities.

Activities suitable for small-group work

Students are likely to work best in small groups if they understand the purpose of the discussion, if they are supported by resources as evidence, if there is no one correct answer, and if some kind of outcome is expected, e.g. a presentation to the rest of the class, a list in rank order, a list of reasons. The kinds of activities which are appropriate are those in which learning would be enhanced by the sharing of ideas and discussion:

- devising questions to frame an enquiry
- devising questions for a visiting speaker
- a DARTs activity (usually carried out in pairs)
- categorising information
- identifying relationships on a concept map
- putting a list into rank order according to importance or preference, e.g. a list of factors, or locations, or priorities for funding
- planning a presentation, e.g. for the whole class, for a role in a role-play, for a television or radio programme
- devising a plan, e.g. for use of a piece of land
- choosing one or two out of several possibilities on the basis of evidence, e.g. location for a production or event, for receiving charity funds
- discussing whether a proposed change or development should go ahead.

Strategies for organising activities

There is a range of strategies for organising group work:

- *Buzz groups.* Students discuss something in pairs for one or two minutes prior to class discussion. This can give students more confidence to contribute to whole-class discussion.
- *Snowballing.* Students start the activity in pairs. Then each pair joins another pair to share what they have done. Then each group of four joins another group of four to share. Finally, discussion is open to the whole class.
- *Jigsaw groups.* This takes place in four stages:
  1. Students start working in a 'home group' with a common interest, e.g. representing a town with traffic problems. They study the data provided about 'their town' and the nature of the problem.
  2. Each home group sends a representative to each of the newly formed 'topic groups' with a particular focus, e.g. different options for solving traffic problems. Data is provided for the topic group to study and discuss. Each representative tries to understand as much as possible about the option.
  3. The representatives reconstitute their home groups and inform each other about the options. The home group decides what would be best for their town.
  4. Each home group presents and explains its decision to the class.
- *Listening triads.* Students work in groups of three each with a specific role: as speaker; as questioner; or as recorder. The speaker talks about something that has been studied. The questioner listens, and asks questions to ensure understanding. The recorder makes notes and provides feedback to the other two. If the work is organised so that there are three aspects to discuss, then students can experience all three roles (Example 4).

Example 4: Small-group work: listening triads. Source: Chris Holt.

In this lesson, listening triads were used to increase students' understanding of what happened in the Kobe earthquake and why.

Starter

- The whole class watches extracts from television news programmes of the Kobe earthquake, which occurred in January 1995.

Development: whole-class preparation for questioning

- The teacher explains that they are going to discuss this in groups of three and that the skills of questioning are important.
- Introduction to open and closed questions with some examples.
- Everyone writes down three open questions about the Kobe earthquake
- Pupils are put into groups of three (listening triads) and numbered 1, 2 and 3.

  Person 1 is going to be the questioner

  Person 2 is going to be the interviewee

  Person 3 is going to be the secretary

Listening triads stage 1

Person 1 questions person 2. The secretary observes and takes notes.

Listening triads stage 2

Each secretary moves to a different group and tells the new group what he/she has written down. The three students in the new group now discuss similarities and differences between what was asked and said in each group.

Plenary debriefing

- Which questions worked best for finding out information? Why?
- Which were your most unsuccessful questions? Why?
- What differences were there between groups in what you found out?
- What did the questioners find easy and difficult?
- What did the interviewees find easy and difficult?
- What did the secretaries find easy and difficult?
- What do you think was the most important information about Kobe? Why?

ENQUIRY WORK IN WHICH SPEAKING AND LISTENING ARE DOMINANT

Role play

In the world outside the classroom, decisions about issues studied in geography are made in public meetings. Role-plays of public meetings provide an excellent format for investigating issues through speaking and listening. Well-managed role-plays give students a far greater proportion of classroom talk than usual and enable a greater range of students to succeed. Quiet students who are reluctant to contribute information or ideas in class gain confidence under the guise of a different persona. Some students who find writing difficult demonstrate considerable skills in presenting, questioning, reasoning and debating. Often those who are relatively quiet in the role-play itself can demonstrate their thinking involvement by their body language. There is a great potential for high levels of engagement in role-play lessons and for discussion of the issues to continue after the lesson outside the classroom.

More detailed discussion of role-play lessons and the types of issues for which they are suitable can be found in Chapter 11.

Hot seating

Hot seating is a strategy in which students, in role, are questioned by other students. The activity is described in detail in Chapter 11 (see page 144).

Press conference

An effective way of using visiting speakers is by inviting them to answer students' questions instead of making a presentation (Pomeroy, 1991). This has the advantages that the visitor is more likely to provide information which is relevant to the students, and that the students are actively engaged in the process of finding out information. The lesson can be set up as a simulated 'press conference' to be followed by writing a report of the visit. The press conference strategy works best

Example 5: Pomeroy's procedure for using visiting speakers.
Source: Pomeroy, 1991.

1. **Briefing.** Students are given information about the visiting speaker, e.g. who it is going to be, from where, when, for how long, and what special knowledge they might be expected to have. Students are introduced to the press conference simulation, the procedure for the press conference and the nature of the report they will be expected to write afterwards (length, audience).

2. **Preparing questions.** One lesson is spent preparing questions. This can be done using a thought shower, sharing, discussing categories of questions, deciding between questions, and discussing a possible sequence. Each student writes a list of possible questions on a sheet of paper and hands it in at the end of the lesson.

3. **Homework for the teacher.** Either the teacher selects one question from each student and highlights it on the sheet, or the teacher selects suitable questions and types them out with the names of questioners, in the agreed order.

4. Students are either handed their questions back with their own question highlighted, or they are handed a complete list of questions.

5. The visitor is introduced and welcomed and questions are asked, including supplementary questions if appropriate and if there is time. Students make some notes.

6. If time permits, students can compare their notes in groups and add to them.

7. Students write their reports.

if it is well integrated into the unit of study and if students have already studied the topic on which the visitor can answer questions. A possible procedure is set out in Example 5. Pomeroy (1991), who used this strategy with a black South African visitor, found that students were highly motivated by the responsibility they had, found that it increased their knowledge and understanding of South Africa and that all had found it enjoyable and memorable.

A similar strategy for visiting speakers is highly successful in the PSHE lessons referred to in Example 1. Using visiting speakers in this way involves students in all aspects of the enquiry process. They devise questions based on their knowledge and understanding. They use a visiting speaker as data, and they usually become aware that the information they hear is from a point of view. They make sense of the information through listening, through further questioning and through trying to represent the findings in a report.

There are several possible sources of visiting speakers:

- other members of staff or student teachers with particular experiences of place or of the topic being studied
- various organisations and businesses
- students from overseas who are studying locally and who can sometimes be contacted through local universities
- parents and other people connected to the school.

SUMMARY

Speaking and listening are crucially important in geography classrooms, both to enable students and teacher to communicate well with each other, and also to enable students to develop their own understanding. Although research has emphasised the importance of speaking and listening in learning, oracy has been relatively neglected in practice since the introduction of the national curriculum.

For students to learn through classroom talk, it is important to establish a classroom environment in which discussion can take place confidently. This can be achieved through negotiating ground rules. In whole-class discussion the kinds of talk that best support learning through geographical enquiry are those in which students have the opportunity to question, speculate, reason and reflect and in which teachers listen and respond to what they say.

Well managed small-group discussion can enable all students in a class to participate in exploratory talk in which they can relate new knowledge to what they know and suggest tentative thoughts and opinions.

Enquiry work in which the main activities are speaking and listening, such as simulated public meeting role-plays and press conferences, are extremely motivating to students and enhance understanding.

"'Twenty-six and five,
thirty-one. Phew! So that
makes five hundred and one
million, six hundred, and
twenty-two thousand, seven
hundred and thirty-one."

"Five hundred million what?"

"Eh? Are you still here? Five
hundred and one million …
erm … I forget. I've so much
work to do! I'm a serious
man, I am. I don't idle
away my life. Two and
five, seven …"

"Five hundred and one
million what?" repeated the
little prince who had never in
his life given up a question
once he'd asked it'

(de Saint-Exupery,
1942, p. 50).

INTRODUCTION

A quick glance at any key stage 3 geography textbook will indicate what a numerate subject geography is. Practically every section of every textbook, regardless of its content, includes numbers. Even where there is a qualitative approach to a theme, the numbers and mathematical concepts keep creeping in. Students need to be able to understand the meaning of, for example: millions of people; percentage; the Richter scale; degrees Celsius, etc. In addition, maps and graphs are an integral part of most geographical enquiry work; students construct them to represent data visually and they interpret them as secondary data. In order to use maps and graphs, students need some understanding of the concepts of co-ordinates and scale.

The mathematical demands of geography, even at key stage 3, are challenging but the work of geography teachers is supported considerably by what students are taught as part of the key stage 3 mathematics curriculum. This chapter presents and discusses aspects of the mathematics curriculum that are particularly relevant to the study of geography. It then introduces some ways in which geography teachers can help to develop the mathematical understanding required for enquiry work.

THE MATHEMATICS NATIONAL CURRICULUM AND THE FRAMEWORK FOR TEACHING

Geography teachers should be aware of three official documents which are available in all schools in England:

- DfEE (1999c) *Mathematics: The national curriculum for England.* This document includes the legal requirements of the national curriculum for mathematics. It sets out, in the programmes of study, what should be taught at each key stage, and sets out in four separate attainment targets what 'pupils' are expected to achieve. This document became statutory in England in September 2000.
- DfEE (2001b) *Key Stage 3 National Strategy: Framework for teaching mathematics: Years 7, 8 and 9.* This document builds on the numeracy strategy for key stages 1 and 2 and aims to provide support for teachers of mathematics. Although the framework is not statutory, schools are expected to use it 'or to justify not doing so by reference to what they are doing' (DfEE, 2001b, p. 2). So it is likely to be used in the majority of schools. Whereas the national curriculum for mathematics sets out *what* should be taught, the framework provides guidance on *when* different aspects of the mathematics national curriculum should be taught and *how* they should be taught. Schools started to implement this framework from September 2001.
- DfES (2001) *Key Stage 3 National Strategy. Numeracy across the curriculum: Notes for school-based training.* These materials were introduced in 2001-02 and secondary schools are expected to have plans for developing numeracy across the curriculum.

These documents indicate the extent of central control of the mathematics curriculum; there is far more detailed prescription than in the geography curriculum. This could be an advantage for geography teachers when they are planning their courses; they can find out from the documentation, as well as through liaison with the mathematics department, when students are taught mathematical skills and concepts and take this into account in curriculum planning. The documents listed above, however, are large, daunting to read, and only some parts are relevant to teaching and learning in geography. So this chapter identifies aspects of the key stage 3 mathematics curriculum that seem to be of particular importance in geographical enquiry, presents selected parts of these documents for reference and discusses some of the issues arising. The four aspects identified are:

- number
- the data-handling cycle as a whole

- using graphs as part of the data-handling cycle
- other miscellaneous mathematical concepts.

NUMBER

At key stage 2 children are introduced in mathematics to whole numbers, including negative numbers (numbers less than 0), fractions and decimals. They are often taught through the use of number lines, measuring scales and long counting sticks, so that they can see the relationship between different numbers. It would be useful for geography teachers to use similar methods to reinforce numerical understanding.

At key stage 3 in mathematics, students are introduced to a wide range of units of measurement (Figure 1). They are introduced to both metric and imperial units and they learn to measure them and read measurements from scales. They are often introduced to negative numbers with reference to temperature in degrees Celsius or with reference to metres below sea level.

The extent to which numbers are easy to understand varies considerably. Generally, numbers are more difficult for students to read and to understand if they are very large or very small or if they include fractions and decimals. Compound measures (numbers which have been calculated from two sets of numbers) are more difficult to understand than simple measures calculated on one linear scale. For example, figures for density of population, calculated from figures for both population and area, are more difficult to understand than figures for distance.

Figure 1: Mathematical vocabulary associated with number. Source: DfEE, 2001b.

Year 7

Area: square millimetre, centimetre, metre, kilometre

Capacity: millilitre, centilitre, litre, pint, gallons

Length: millimetre, centimetre, metre, kilometre, mile

Mass: gram, kilogram, ounce, pound

Time: second, minute, hour, day, week, month, year, decade, century, millennium

Temperature: degrees Celsius, degrees Fahrenheit

Depth

Distance

Height, high

Perimeter

Surface, surface area

Width

Year 8

Foot, yard

Hectare

Tonne

Volume: cubic millimetre, cubic centimetre, cubic metre

Year 9

Density

Pressure

Speed: kilometres per hour, metres per second

THE DATA-HANDLING CYCLE

The clearest links between geographical enquiry and mathematics come under the heading of 'data handling', which is allocated its own substantial section, Ma4, in the mathematics national curriculum programme of study, and its own attainment target, AT4. Data handling is envisaged as a cyclical process (Figure 2) with four stages (Figure 3). This all looks remarkably familiar!

Furthermore, the mathematics framework for key stage 3 emphasises the importance of enquiry in mathematics education:

'Enquiry lies at the heart of mathematics. Enquiry skills enable pupils to ask questions, define questions for enquiry, plan research, predict outcomes, anticipate consequences and draw conclusions. Central to enquiry is an ability to see connections between different aspects of mathematics and thus open up further ways of tackling a problem' (DfEE, 2001b, p. 21).

The Framework provides guidance on how and when these enquiry skills should be taught:

'Data handling is best taught in a coherent way in the context of real statistical enquiries so that teaching objectives arise naturally from the whole cycle. It is easier to make sure that problems are relevant if at least some of the enquiries are linked to other subjects. For example, a question can be formulated and data collected in geography, with mathematics lessons concentrating on processing, representing and

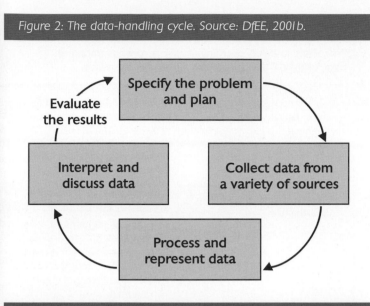

Figure 2: The data-handling cycle. Source: DfEE, 2001b.

Figure 3: Four aspects of data handling in mathematics. Note: Further details of what students should be taught in each of the four aspects of the data-handling cycle listed in section (a) are set out on pages 40-41 of DfEE, 1999c. Source: DfEE, 1999c, p. 40.

'Pupils should be taught to:

a. Carry out each of the four aspects of the handling cycle to solve problems:

  i. Specify the problem and plan; formulate questions in terms of the data needed, and consider what inferences can be drawn from the data; decide what data to collect (including sample size and data format) and what statistical analysis is needed.

  ii. Collect data from a variety of suitable sources, including experiments and survey, and primary and secondary sources.

  iii. Process and represent the data; turn the raw data into usable information that gives insight into the problem.

  iv. Interpret and discuss the data; answer the initial question by drawing conclusions from the data.

b. Identify what further information is required to pursue a particular line of enquiry.

c. Select and organise the appropriate mathematics and resources to use for a task.

d. Review progress as they work: check and evaluate solutions.'

interpreting the data. The other subject can make further interpretations and consider the implications of what has been discovered' (DfEE, 2001b, p. 20).

The mathematics curriculum is emphasising here what is so important in geographical enquiry: skills are developed not as ends in themselves but in relevant, purposeful contexts. It is satisfying that geography is quoted as a provider of such relevance and that liaison between mathematics and geography teachers is encouraged. Geography departments could indeed suggest real contexts for data-handling work. Mathematics departments could support geography departments by providing advice on how teachers can develop or reinforce what has already been taught in maths. It is, however, questionable whether the division of labour suggested above is the most appropriate way of working. It interrupts the coherence of the enquiry cycle and might focus attention on separate data-handling skills, rather than the purpose for which they are being used. Both subjects can contribute in important ways to each stage of the data handling cycle. Data collection involves mathematical decisions as well as geographical skills. Data representation and interpretation involves geographical judgements as well as mathematical skills. Mathematics and geography lessons provide students with different perspectives on the decisions that are made in data handling and contribute to the development of different but complementary skills. Figure 4, based on the key stage 3 framework, shows when the complete range of data-handling skills is likely to be developed in mathematics.

USING GRAPHS

One part of the data-handling cycle involves the construction and interpretation of graphs. It is useful for geography teachers to know what types of graph are taught in mathematics, what students are expected to be able to do with them, and when they are taught. This information, based on the mathematics national curriculum and on the *Key Stage 3 Framework for Teaching Mathematics* (DfEE, 2001b), is summarised in Figure 5. It demonstrates that most of the types of graph used in geography are taught in mathematics. There are, however, some issues relating to pie charts, scatter graphs and climate graphs.

In mathematics, students are not expected to draw pie charts by hand until year 8. Figures 4 and 5 indicate that they will have been introduced to simple pie charts in year 6 and that in year 7 they are expected to interpret pie charts, to compare them and to construct them using a computer.

Year	Specifying a problem, planning and collecting data	Processing and representing data, using ICT as appropriate	Interpreting data and discussing results
	Figure 4: Data handling: guidance on what should be taught in each year in mathematics. Source: DfEE, 2001b, pp. 7-11.		
7	• Given a problem that can be addressed by statistical methods suggest possible answers. • Decide which data would be relevant to the enquiry or identify possible sources. • Plan how to collect and organise small sets of data; design a data collection sheet or questionnaire to use in a simple survey; construct frequency tables for discrete data, grouped where appropriate in equal class intervals. • Collect small sets of data from surveys and experiments, as planned.	• Calculate statistics for small sets of discrete data: – find the mode, median and range – calculate the mean. • Construct, on paper and using ICT, graphs and diagrams to represent data (see Figure 5).	• Interpret diagrams and graphs and draw simple conclusions based on the shape of graphs and simple statistics for a single distribution. • Compare two simple distributions. • Write a short report of a statistical enquiry and illustrate with appropriate diagrams, graphs and charts, using ICT as appropriate, and justify the choice of what is presented.
8	• Discuss a problem that can be solved by statistical methods and identify related questions to explore. • Decide which data to collect to answer a question and the degree of accuracy needed; identify possible sources. • Plan how to collect the data, including sample size; construct frequency tables with given equal class intervals for sets of continuous data; design and use two-way tables for discrete data. • Collect data using a suitable method, such as observation, controlled experiment, including data logging using ICT, or questionnaire.	• Calculate statistics; recognise when it is appropriate to use the mean, median and mode. • Construct, on paper and using ICT, graphs and diagrams to represent data (see Figure 5) and identify which are most useful in the context of the problem.	• Interpret tables, graphs and diagrams. • Compare two distributions. • Communicate orally and on paper the results of a statistical enquiry and the methods used; justify the choice of what is presented.
9	• Suggest a problem to explore using statistical methods, frame questions and raise conjectures. • Discuss how data relate to the enquiry; identify possible sources, including primary and secondary sources. • Design a survey or experiment to capture relevant data from one or more sources; determine the sample size and degree of accuracy needed; design, trial and if necessary refine data collection sheets; construct tables for large discrete and continuous sets of raw data, choosing suitable class intervals; design and use two-way tables. • Gather data from specified secondary sources including printed tables and lists from ICT based sources.	• Select the statistics most appropriate to the problem. • Select, construct and modify, on paper and using ICT, suitable graphical representation to progress an enquiry (see Figure 5) and identify key features present in the data.	• Interpret graphs and diagrams and draw inferences to support or cast doubt on initial conjectures; have a basic understanding of correlation. • Compare two or more distributions and make inferences, using the shape of the distributions, the range of data and appropriate statistics. • Communicate interpretations and results of a statistical enquiry using selected tables, graphs and diagrams in support using ICT as appropriate.

The construction of pie charts by hand is difficult; it involves the translation of proportions or percentages into angles and the use of a protractor. Although students are introduced to angles in year 5 and to the use of protractors to measure and construct angles in year 6, many still find their use difficult in year 7. If geography teachers want their year 7 students to construct pie charts by hand, it would be preferable to provide them with a skeleton outline on which percentages are marked, from which students can estimate where to mark the segments (Figure 6).

Scatter graphs are not introduced in the *Framework* (DfEE, 2001b) until year 8, with the concept of best-fit line being introduced only to 'more able' students in year 9. Scatter graphs are difficult to understand for a variety of reasons. They involve the use of continuous data which are more difficult to understand than categorical or discreet data. They involve the use of two sets of data. Very often scatter graphs display the relationship between compound measures, such as density and percentage, rather than simple measures. The pattern of what is plotted is a hypothetical relationship. Because of all this, the reading and interpretation of scatter graphs requires considerable mathematical understanding. It would be wise for geography teachers to delay their use until year 8 or until students have good understanding of continuous data and compound measures.

One puzzling aspect of the *Framework* is its advice on climate graphs. Both temperature and rainfall figures are plotted with lines. Students are expected to 'know that it can be appropriate to join the points on the graphs in order to compare trends over time' (DfEE, 2001b, p. 265). This advice contradicts the way climate graphs are conventionally constructed not only in geography textbooks and atlases, but also by meteorologists. The convention is that temperature is plotted with a line and rainfall is plotted with bar graphs. There is a good reason for this convention; mean monthly temperature and mean rainfall figures are calculated in different ways. Mean monthly temperature figures are calculated by:

Figure 5: Graphs included in the key stage 3 mathematics national curriculum: types, the year they are taught, and assessment levels.		
Bar-line graph	**Bar chart for categorical data**	**Bar chart for grouped discrete data**
Used for: discrete variables, where the length of bar represents the frequency. It is not appropriate to join the tops of the bars. Y5: represent and interpret Y6: construct and interpret	Suitable for data which can be allocated to distinct categories. The bars are labelled with the category. AT Level 3: construct and interpret Y6: construct, interpret *Categorical data* is data which can be allocated to distinct categories, e.g. ways of travelling to school.	Suitable for data that can be allocated into categories using suitable class intervals, e.g. the bars are labelled with the range they represent, but not the divisions between the bars. Y7: choosing suitable class intervals, construct, interpret
Compound bar chart	**Frequency diagram for a continuous variable**	**Population pyramid**
Compound bar charts allow both overall trends and changes in subcategories to be shown. Y7: interpret data Y8: construct on paper and using ICT, interpret	Frequency diagrams are suitable for continuous variables, e.g. time. For continuous data the divisions between the bars should be labelled. AT level 4: represent data Y8: construct, interpret (level 6) Y9: construct, interpret	Y8: interpret Y9: interpret
Pie chart	**Line graph**	**Scatter graph**
Suitable mainly for categorical data. AT level 5: interpret AT level 6: construct Y6: interpret simple pie charts Y7: interpret, construct using ICT Y8: construct on paper and using ICT	AT level 4: construct and interpret Y5: construct and interpret Y9: construct and interpret, using more than two sets of data	Scatter graphs are used for continuous data, with two variables plotted against each other. AT level 6: draw conclusions from scatter graphs AT level 7: draw a line of best fit on a scatter graph Y8: construct and interpret Y9: construct, interpret, draw a line of best fit by eye or using spreadsheet

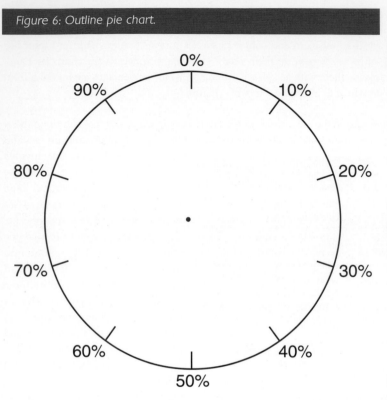

Figure 6: Outline pie chart.

- finding the mean figure for each day of the month (by adding the maximum and minimum temperatures and dividing by two)
- finding the mean figure for a particular month in a particular year (by adding the daily temperature means and dividing by the number of days in the month)
- finding the mean figure for each month over a period of 35 years (e.g. by adding the mean January temperature figures for each of the 35 years and dividing by 35).

Whereas mean monthly rainfall figures are calculated by:

- finding the total rainfall for a particular month in a particular year (by adding the amount of rain collected each day)
- finding the mean figure for each month over a period of 35 years (e.g. by adding the total figures for January for each of the 35 years and dividing by 35).

Except when figures are plotted for one particular year, both temperature and rainfall figures on climate graphs are *mean* figures, in spite of what is written in the *Numeracy Across the Curriculum* Notes (DfES, 2001). However, because the temperature figures have been calculated from daily mean figures, the line on the graph does indeed show a trend and a reading at any point along the line has some significance. The rainfall figures have been calculated from monthly totals, and readings on a trend line serve no useful purpose. It would be helpful for geography teachers to discuss the conventions related to climate graphs with mathematics colleagues.

The *Framework* (DfEE, 2001b) includes lists of vocabulary that students are expected to 'use, read and write, spelling correctly'. The vocabulary used in mathematics associated with data handling, including graphs, is listed in Figure 7. It is important for geography teachers to use the same vocabulary as mathematics teachers to help students develop an appropriate and consistent language for use in reference to data handling.

OTHER CONCEPTS

Other miscellaneous concepts relevant to geography that are taught in mathematics at key stage 3 include:

- the use of co-ordinates
- ratio and proportion
- probability.

Students are introduced to co-ordinates in year 5 and learn to read and plot points in the first quadrant in year 5 and in all four quadrants in year 6 (Figure 8). In the first quadrant, points are located in reference to the x-axis below the quadrant and the y-axis to the left of the quadrant. Grid references are located in a similar way and to be consistent with mathematics it is best for geography teachers to encourage

Figure 7: Vocabulary associated with data handling. Source: DfEE, 2001b.

Year	Vocabulary		
7	Average	Frequency diagram	Range
	Bar chart	Interpret	Represent
	Bar-line graph	Interval	Statistics
	Class interval	Label	Survey
	Data, grouped data	Mean	Table
	Data collection sheet	Median	Tally
	Experiment	Mode	Title
	Frequency	Pie chart	
	Frequency chart	Questionnaire	
8	Continuous	Interrogate	Scatter graph
	Data log	Line graph	Secondary source
	Discrete	Population pyramid	Stem and leaf diagram
	Distance-time graph	Primary source	
	Distribution	Sample	
9	Two-way table	Estimate of the mean/median	Quartiles
	Bias		Raw data representative
	Census	Interquartile range	
	Cumulative frequency	Line of best fit	

Figure 8: Co-ordinates: the four quadrants. Source: DfEE, 2001b.

Students should be taught to use co-ordinates in all four quadrants. As outcomes, year 7 students should, for example:

Use, read and write, spelling correctly:
Row, column, co-ordinates, origin, x-axis, y-axis, position, direction, intersecting, intersection.

Read and plot points using co-ordinates in all four quadrants

y-axis

Second quadrant | First quadrant

x-axis

Third quadrant | Fourth quadrant

students always to refer to the grid figures below and to the left of the map. The ability to use latitude and longitude to read and plot points on a map is more complicated and depends on being able to use all four quadrants; students need to be able to read figures in two directions (east and west) starting at the Greenwich (or Prime) Meridian and in two directions (north and south) starting at the Equator. This can be explicitly compared with reading co-ordinates in all four quadrants.

Students are first introduced to simple fractions in year 2 and by year 5 are taught to understand percentage as 'the number of parts in every 100' (DfEE, 2001b, p. 100). This understanding becomes a core objective in year 6 mathematics; geography teachers can reinforce this understanding by using this same phrase: 'the number of parts in every 100'. Students begin to learn about ratio in year 7 and are taught in mathematics how it is expressed in a range of contexts including scales on maps, e.g. 1:10,000.

The language of probability is introduced in year 5 and is further developed in years 6 and 7. Students are taught to use specific vocabulary:

- certain
- impossible
- likely
- unlikely
- even chance.

This vocabulary can be reinforced through its use in speculative discussion in geography.

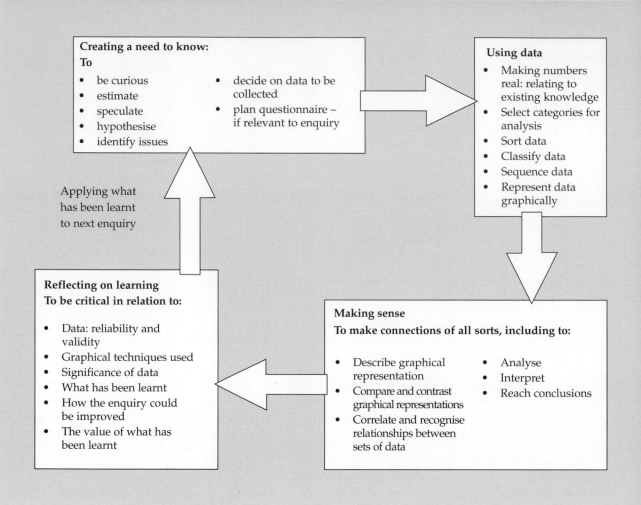

Figure 9: A framework for learning through enquiry with the focus on numeracy.

Creating a need to know:
To
- be curious
- estimate
- speculate
- hypothesise
- identify issues
- decide on data to be collected
- plan questionnaire – if relevant to enquiry

Using data
- Making numbers real: relating to existing knowledge
- Select categories for analysis
- Sort data
- Classify data
- Sequence data
- Represent data graphically

Applying what has been learnt to next enquiry

Reflecting on learning
To be critical in relation to:
- Data: reliability and validity
- Graphical techniques used
- Significance of data
- What has been learnt
- How the enquiry could be improved
- The value of what has been learnt

Making sense
To make connections of all sorts, including to:
- Describe graphical representation
- Compare and contrast graphical representations
- Correlate and recognise relationships between sets of data
- Analyse
- Interpret
- Reach conclusions

ACTIVITIES RELATED TO NUMERACY

In geographical enquiry work it is important that the use of numbers enhances the understanding of geography at the same time as contributing to the development of numeracy. The activities discussed below are all envisaged as an integral part of the enquiry process (Figure 9):

- making numbers real: relating to experience; high/low charts; common knowledge
- intelligent guesswork
- best and worst
- active numbers
- other activities: categorising data; labelling graphs or maps; talking graphs or maps; questionnaire survey

Example 1: High/low indices of development.

Gross Domestic Product per capita	
High GDP per capita	34,142 US $ per person
World average GDP per capita	7,446 US $ per person
Low GDP per capita	490 US $ per person

MAKING NUMBERS REAL

Relating numbers to students' experience

Geography teachers need to make numbers real whenever they are used. There is no point in expecting students to use data about, for example, hectares, or GNP (gross national product) or death rates per thousand if they have no idea what the figures actually represent in the real world. It is all too easy to assume that because students have a definition in words, they really understand. Making numbers real involves relating numbers to students' existing knowledge and experience or to what they can envisage or imagine, using examples, illustrations or analogies. For example, numbers of people can be related to the number of people in a class, in the school, in the town or city where the school is situated or to the capacity of a sports stadium, e.g. the Millennium Stadium in Cardiff has a capacity of 75,000. Distances can be explained in terms of distances from the school to particular places and can be referred to both in kilometres and in time taken to travel this distance using different forms of transport. Small areas can be related to football pitches or other sports pitches or courts (the average football pitch is half a hectare) and the size of other countries can be related to the size of the UK or Britain.

It is particularly important to make numbers real and to encourage accuracy when discussing percentages, rates per thousand and compound measures. A very common error is for students to use the word 'more' when they are referring to statistics or graphs which indicate a greater percentage or a higher rate per thousand. The total figure might well not be more. Geography teachers may need to draw particular attention to the meaning of rates per 1000 which may not be encountered in mathematics but which are always used for birth and death rates.

High/low charts

Students may understand the meaning of numbers and units of measurement, but not know whether the particular figures they encounter are average, high, or low. A useful way of emphasising the range of numbers associated with a particular measurement, e.g. average calories per capita, is to produce high/low display charts, preferably illustrated with some visual representation for emphasis. Example 1 shows one of several A4 sheets produced for display during the study of indices of development.

Making numbers real as common knowledge

In order to 'make numbers real' teachers need to get inside students' minds. They need to know what students are likely to know, understand and have encountered, so that they can use examples that will mean something to the students. Classroom dialogues to 'make numbers real' can become part of the 'common knowledge' of the class, points of reference for future discussions. Making numbers real can be developed as a starter activity but numbers need to be real to students whenever they are being used.

INTELLIGENT GUESSWORK

Intelligent guesswork is an activity designed to 'create a need to know'. In this activity students make guesses related to statistical information before they are actually given the full information. Most students have some awareness of many aspects of geography, gained from their own experiences, from other people, the media, and from formal education. There is a wide range of topics on which they might be expected to make intelligent guesses, for example:

- population, e.g. life expectancy figures, total population figures of countries, population density figures, migration figures
- tourism, e.g. most frequently visited tourist attractions in England; most visited EU countries, countries of origin of tourists to the UK
- weather and climate, e.g. maximum and minimum temperatures of different places, places of origin of climate graphs
- use of water, e.g. how much water is used for different purposes in UK households
- use of energy in UK, e.g. what proportion of energy comes from different sources.

Of course, intelligent guesswork can be used only if there is a chance that students can make a reasoned estimate. For example, students might reasonably be expected to estimate which countries in Europe receive most visitors from the UK but not to estimate the population density of different states in India.

Example 2: Intelligent guesswork.

(a) Procedure

1. Students are introduced to the focus of the enquiry work. They are provided with information with some element missing, e.g. lists of countries without figures for particular indices. The teacher needs to adopt a stance towards the information which creates an element of puzzle and which generates curiosity (e.g. I wonder whether…? Could the figure for x be something like … I wonder? I am really interested to know what guesses you will make).

2. Students work individually or in pairs. They make intelligent guesses, estimating numbers or rank order, and write them down.

3. Whole class discussion: sharing of the guesses. The teacher accepts all guesses without commenting on whether they are right or wrong. The teacher adopts a curious stance (I wonder if that could be right? What do others think?).

4. Whole class discussion: probing the thinking. The teacher probes the thinking behind the guesses (that is interesting, what made you choose that number/place? How sure are you about that? What do other people think?). Reasons behind the thinking can be summarised on the board. What is happening in this part of the lesson is that the students are revealing their knowledge and the way they reason geographically. Misconceptions often emerge and these can be picked up later, when the figures are revealed or later in the enquiry work.

5. Provision of answers: The teacher asks the class whether they want to know the 'answers'. (They always do!) The teacher provides the data and students write this down next to their guesses.

6. Debriefing: Which figures did you get right? Why do you think this was? What figures were surprising? Why? Which of your explanations for these numbers seem most likely?

(b) Task sheet

For each country in the table: Estimate the average life expectancy at birth and write down the figure in column 2.

- Work out, according to your estimates, the rank order of these 16 countries for average life expectancy. In column 3, put 1 against the country with the highest life expectancy, 2 against the country with the next highest, etc.
- Underline the names of the countries ranked 1 and 2.
- Underline in a different colour the names of the countries ranked 15 and 16.

1 Name of country	2 Life expectancy at birth	3 Rank
Australia		
Bangladesh		
Bolivia		
Brazil		
China		
India		
Italy		
Jamaica		
Japan		
Malawi		
Poland		
Saudi Arabia		
South Africa		
Tunisia		
UK		
USA		

Example 2: Intelligent guesswork ... continued

(c) Life expectancy at birth: data for 16 countries.
Source: http://hdr.undp.org/hdr2006/statistics/indicators/2.html

1 Name of country	2 Life expectancy at birth (guess)	3 Rank	4 Actual life expectancy
Australia	80	=2	81
Bangladesh	59	11	63
Bolivia	63	10	64
Brazil	69	=9	71
China	71	8	72
India	69	=9	64
Italy	80	=2	80
Jamaica	75	5	71
Japan	81	1	82
Malawi	38	13	40
Poland	74	6	75
Saudi Arabia	72	=7	72
South Africa	51	12	47
Tunisia	72	=7	74
UK	78	3	79
USA	77	4	78

Intelligent guesswork is a valuable activity for several reasons:

1. It can be very motivating. Students can develop a keen need to know the information.

2. Discussion of students' guesses can reveal the way they think geographically.

3. If the reasons for the information are discussed, students become motivated to know about explanations for the information.

4. It draws attention to particular information related to the topic of study, making it more memorable.

Intelligent guesswork is a starter activity which needs to be followed by activities in which some sense is made of the data used. A general procedure for using the strategy of intelligent guesswork is set out in Example 2a.

Example 2b is an intelligent guesswork worksheet on which students estimate life expectancy in selected countries and then complete the rank order column for this set of countries. The 'answers', from statistics published in 2007, are shown in Example 2c. In using this example, it is worth concentrating the discussion on which countries have been estimated to have the highest and lowest life expectancies. The thinking behind these estimates can reveal a lot about students' understanding and reasoning. This activity and worksheet can easily be adapted for other topics.

BEST AND WORST

Which are the best places to live and which are the worst? In this activity students study numerical data to determine which would be the best place to live and which would be the worst place to live. The activity could be carried out in relation to:

- towns (e.g. in the UK), regions (e.g. of Europe) or countries
- a particular group of people, e.g. children, women (e.g. Which is the worst country in the world for children under 10?)
- published statistics on indices of development, or could be related to distances from certain facilities, e.g. distance of a town from a Premier Division football club.

A possible procedure is set out in Example 3.

ACTIVE NUMBERS

Another way of making numbers real is to act them out in the classroom. The psychologist Jerome Bruner (1966) wrote about three different ways in which knowledge can be presented:

- formal representation: through words or numbers
- iconic representation: through the use of summarising visual images such as pictures, graphs, maps and icons
- enactive representation: through action.

Example 3: Best and worst.

(a) A suggested procedure

Starter: creating a need to know

Initial thought shower:

- If you had to live in another town/another country where would you like to live?
- Why would these places be good places to live?
- I wonder if you would agree with lists of places chosen by organisations, e.g. United Nations (Example 3b)?

Using data: deciding criteria

- Students study a list of criteria and decide which five criteria are the most important for them.

Making sense of data: ranking

This part of the enquiry could be carried out using information on a database on a computer or could be based on the use of statistics, atlases or maps.

Students' task is to:

- rank a given list of places according to each of the criteria
- add up the rank numbers to achieve a total figure
- identify their 'best and worst' places
- locate their 'best and worst' places on a map
- annotate the map with details of their 'best and worst' places

Reflecting on learning

Debriefing:

- Which places have emerged as best and worst (locate on large class map if possible)?

- If there is another ranking of these places (e.g. by the United Nations) how does the ranking compare? (Reveal other rankings.)
- Is there a pattern in the distribution of best and worst places?
- Are the figures about these places the whole story? Are there things that might make these best places not so good to live in? Are there things that might make the worst places not so bad?

(b) An example

Rank order of the best and worst countries in the world to live in according to the Human Development Index (based on life expectancy, education and GDP).

'Best' 10 countries		'Worst' ten countries	
1	Norway	164	Mali
2	Sweden	165	Central African Republic
3	Canada	166	Chad
4	Belgium	167	Guinea-Bissau
5	Australia	168	Ethiopia
6	United States	169	Burkina Faso
7	Iceland	170	Mozambique
8	Netherlands	171	Burundi
9	Japan	172	Niger
10	Finland	173	Sierra Leone

Source: www.undp.org/hdr2002 and www.infoplease.com/ipa/A0778562.html

Bruner argued that the most intellectually demanding form of representing knowledge was symbolically through language and numbers. It was also the most powerful: words and numbers could be used for manipulating ideas, forming hypotheses and for representing thought in condensed and powerful ways. If we want to make ideas easier to understand, we can represent them visually. If we want to make them even more accessible, then we can represent them enactively, i.e. through action. Enactive representation is particularly useful for teaching geographical ideas related to numeracy because of the problems some students have in mathematical understanding.

What this means in the classroom is that the numbers are acted out in some way with the students representing the numbers. Enactive representation can be used in a variety of ways ranging from very short activities to whole-lesson enquiries. In Example 4, students learn about the population and distribution of land in South Africa through an active numbers enquiry. If students are to gain any understanding of present-day South Africa they need some knowledge and understanding of South Africa during Apartheid. In this activity it is intended that students experience the process of classification, segregation and different treatment, not simply as facts but as political decisions affecting quality of life. Students learn through action and through memorable experience. This activity can be adapted to apply to other topics investigated in geography.

Example 4: Active numbers: Understanding South Africa.

(a) Instructions

Key questions

- How were the people of South Africa classified during the Apartheid years?
- How were the people of South Africa segregated during the Apartheid years?
- How was the land of South Africa divided during the Apartheid years?
- What should be done about the land issue?

Resources

- Cards, each card representing 1.25 million people (Example 4b)
- An edible pie chart (sponge cake or tart, but not too messy!), knife, paper towels
- One copy of pie chart to show how cake should be cut (Example 4c)
- Labels for 'homelands' to be placed around the margins of classroom to represent the former 'homelands' on the margins of South Africa. (The 'homelands' are listed in the key to the pie chart. For this activity the two homelands of Transkei and Ciskei have been combined.)
- Map of South Africa showing the provinces (Example 4d)
- Information about 'racial' classification (Example 4f)
- Map of South Africa showing the location of the former homelands (Example 4g)

Procedure

Starter: setting the context

Explain that the room represents the whole of South Africa and that students represent all the people of South Africa, a total of approximately 40 million people. (The statistics on which this activity is based are shown in Example 4e.)

Give each student a card (each representing approximately 1.25 million people).

Ask a few students to read the information on their cards aloud.

Classification

Tell students that between 1950 and 1991 everyone in South Africa was classified, within seven days of birth, into one of four 'racial' groups (Example 4f). Details of the four main categories are displayed. Students decide how they would be classified.

- Check which students think they would be classified as Indian (card 30), coloured (cards 24, 25 and 26) and white (cards 27, 28 and 29) and black (the rest of the class).
- When each student is classified correctly, each 'racial' group stands up in turn so that the proportions of different groups in the population of South Africa as a whole can be experienced.

Segregation and removals

Tell students that between 1950 and 1991 all the land in South Africa was allocated to different 'racial' groups and that everyone had to live apart from other groups. Black people were further classified according to their language groups and are allocated to 'homelands' (Examples 4d and 4g). Many people had to move to the areas allocated to them.

- Display and read out the names of the 'homelands' and the first language of the group allocated to it. For example, those whose first language is Zulu are allocated to KwaZulu (check that everyone is looking at the *first* language as many of those classified as black speak several languages).
- Point out where each 'homeland' is on the map and in the classroom. Check, by asking a few individuals, that all those classified as 'black' know where to move. Students move to their 'homelands' on the margins of the classroom and stand near the labels.
- Those classified as 'white' move to a large desk or table at the front.
- Those classified as 'coloured' move to a desk segregated from the 'whites'.
- The person classified as 'Indian' moves to a desk, segregated from the other groups.
- (To make the classroom resemble the Apartheid years more accurately, about 8 of those classified as black could move from their homelands to be nearer the white people, to represent townships, farm workers, etc. These students should sit segregated from the white people.)

Discrimination

During the apartheid years people had rights to own land, property and businesses only in the land allocated to them.

- Show the students the cake, the edible pie chart, which represents all the land of South Africa.
- Divide the cake into segments according to the pie chart and hand out to the groups, starting with the homelands. Those classified as 'Indian' or 'coloured' can be given some crumbs to represent the areas allocated to them in towns. Nobody eats the cake yet!

The end of Apartheid

The laws of Apartheid which classified and segregated people were repealed between 1991 and 1994. It was then possible for people to move where they wanted to and buy property, land and businesses (if they could afford it).

- Ask six of those classified as black to join the white people at their table. The black people represent those who can afford to move into white areas, buy land, etc. They might work, for example, in business, in local or national government, or in the entertainment business. Now, from the point of view of the white people, apartheid has disappeared.
- Most of the students represent those for whom there has been less change. The land (the cake) is still unevenly distributed.
- Students get together in groups and decide what should be done now to make the distribution of the land (and cake) fairer.
- Each group presents its ideas.

Note: In South Africa during the last eight years a small percentage of the land from which black people were removed has now been returned to the original owners. Most, however, is still in white ownership.

Example 4: Active numbers: Understanding South Africa ... continued

(b) Population cards

1 Name: Madumetja Phatudi (M) Address: Seshego, near Pietersburg Province: Northern Province Ancestry: metal workers living on the African plateau First language: North Sotho Other languages: Afrikaans, English	**2** Name: Mogoboya Machaka (M) Address: Nanedi, near Potgeitersrus Province: Northern Province Ancestry: farmers living on the African plateau First language: North Sotho Other languages: Afrikaans	**3** Name: Tsiame Motha (M) Address: Isando, near Johannesburg Province: Gauteng Ancestry: farmers living on the African plateau First language: North Sotho Other languages: Tswana, English
4 Name: Samuel Letsoko (M) Address: Thaba'Nchu, near Bloemfontein Province: Free State Ancestry: farmers living on the African plateau First language: South Sotho Other languages: Afrikaans, English	**5** Name: Mamphuthela Chaka (F) Address: Kroonstadt Province: Free State Ancestry: farmers living on the African plateau First language: South Sotho Other languages: Afrikaans	**6** Name: Grace Sithole (F) Address: Barberton Province: Mpumalanga Ancestry: farmers living on the African plateau First language: Swazi Other languages: English
7 Name: Mosegi Matabogi (M) Address: Ga Rankuwa, near Pretoria Province: North West Ancestry: farmers living on the African plateau First language: Tswana Other languages: English, Afrikaans	**8** Name: Kefilwe Setshedi (F) Address: Babelegi, near Pretoria Province: North West Ancestry: farmers living on the African plateau First language: Tswana Other languages: English, North Sotho, South Sotho	**9** Name: Tale Gaobebe (F) Address: Tembisa, near Johannesburg Province: Gauteng Ancestry: farmers and metal workers living on the African plateau First language: Tswana Other languages: Afrikaans, English
10 Name: Nkateko Mayevu (M) Address: Nelspruit Province: Mpumalanga Ancestry: farmers living on the African plateau First language: Tsonga Other languages: Afrikaans, English, Tswana	**11** Name: Mutheiwana Mphephu (F) Address: Pietersberg Province: Northern Province Ancestry: workers in stone and metal and farmers living on the African plateau First language: Venda Other languages: North Sotho, Afrikaans	**12** Name: George Mniki (M) Address: Umtata Province: Eastern Cape Ancestry: farmers living on the African plateau First language: Xhosa Other languages: English, Zulu
13 Name: Lizile Sebe (M) Address: Lusikisiki Province: Eastern Cape Ancestry: farmers living on the African plateau First language: Xhosa Other languages: English, Zulu, South Sotho	**14** Name: Ncumisa Xuma (F) Address: East London Province: Eastern Cape Ancestry: farmers living on the African plateau First language: Xhosa Other languages: none	**15** Name: Nonceba Ntahele (M) Address: King Williams Town Province: Eastern Cape Ancestry: farmers living on the African plateau First language: Xhosa Other languages: English, Tswana, Afrikaans, Zulu

Example 4: Active numbers: Understanding South Africa ... continued

16	17	18
Name: Thandiwe Tshezi (F) Address: Soweto, near Johannesburg Province: Gauteng Ancestry: farmers living on the African plateau First language: Xhosa Other languages: none	Name: Cuthbert Lembedi (M) Address: Kwa Mashu, near Durban Province: KwaZulu Natal Ancestry: farmers living on the African plateau First Language: Zulu Other languages: Xhosa, English	Name: Siphokazi Sibiya (F) Address: Inanda, near Durban Province: KwaZulu Natal Ancestry: farmers living on the African plateau First language: Zulu Other languages: English
19	**20**	**21**
Name: Constance Ndlovu (F) Address: Edendale, near Pietermaritzburg Province: KwaZulu Natal Ancestry: farmers living on the African plateau First language: Zulu Other languages: English, Afrikaans	Name: Gabriel Khumalo (M) Address: Eshowe Province: KwaZulu Natal Ancestry: farmers living on the African plateau First language: Zulu Other languages: English	Name: Lawrence Mkhizi (M) Address: Vereeniging Province: Gauteng Ancestry: farmers living on the African plateau First language: Zulu Other languages: English, Afrikaans
22	**23**	**24**
Name: Dorothy Mbuyisa (F) Address: Alexandra Township, near Johannesburg Province: Gauteng Ancestry: farmers living on the African plateau First language: Zulu Other languages: English	Name: Ntela Magwaza (M) Address: Port Elizabeth Province: Eastern Cape Ancestry: farmers living on the African plateau First language: Zulu Other languages: English	Name: Polly Botha (F) Address: Khayelitsha, Cape Flats Province: Western Cape Ancestry: Khoi-khoi herders, early Dutch settlers, slaves from Madagascar First language: Afrikaans Other languages: English
25	**26**	**27**
Name: Ismail Vayez (M) Address: Cape town Province: Western Cape Ancestry: Slaves brought from Dutch East Indies in seventeenth century First language: Afrikaans Other languages: English	Name: Stanley Mohr (M) Address: Kimberley Province: Northern Cape Ancestry: San hunters, Dutch settlers, Xhosa people First language: Afrikaans Other languages: English	Name: Jan Engelbrecht (M) Address: Cape Town Province: Western Cape Ancestry: Dutch settlers, English settlers First language: English Other languages: Afrikaans
28	**29**	**30**
Name: Rosemary Rice (F) Address: Durban Province: KwaZulu Natal Ancestry: all ancestors can be traced back to England, including a group from London who arrived in Eastern Cape in 1820s First language: English Other languages: none	Name: Erica de Villier (F) Address: Johannesburg Province: Gauteng Ancestry: settlers who came originally in the seventeenth and eighteenth centuries from The Netherlands and France First language: Afrikaans Other languages: English	Name: John Naidoo (M) Address: Phoenix, near Durban Province: KwaZulu Natal Ancestry: Contract workers who came in 1884 from India to work on sugar plantations First language: English Other languages: Hindi

Example 4: Active numbers: Understanding South Africa ... continued

(c) Land distribution in South Africa during the Apartheid years

(d) Location of places mentioned in South Africa

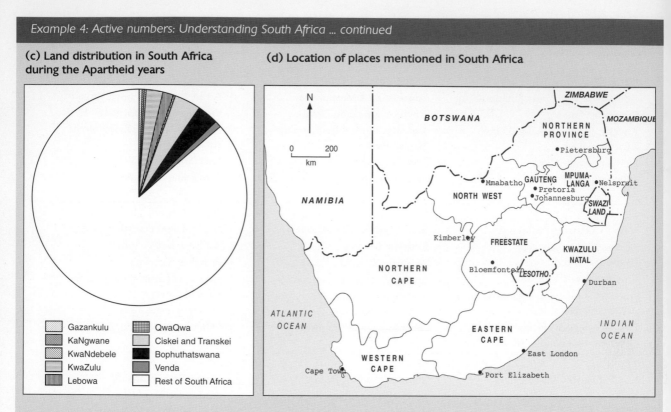

(e) Statistics for South Africa

The cards for this activity (Example 4b) are devised so that the proportion of people in each 'racial group', the proportion of people living in each province, and the distribution of 'racial' groups within South Africa are roughly correct (as far as possible with so few cards).

Classification into 'racial' groups

Classification	Millions of people	Percentage of population	Number of cards
Black	31.0	76.7	23
Coloured	3.6	8.9	3
Indian	1.0	2.6	1
White	4.4	10.9	3
Total for South Africa	40.0	100.0	30

Source: 1996 Census data South Africa

The allocation of language groups to the 'homelands' during the Apartheid years

Name of homeland	First language	Percentage of land area of South Africa
Bophuthatswana	Tswana	3.4
Ciskei and Transkei	Xhosa	4.1
Gazankulu	Tsonga	0.6
Kangwane	Swazi	0.2
Kwandebele	Ndebele	0.1
KwaZulu	Zulu	2.5
Lebowa	North Sotho	1.9
Qwa Qwa	South Sotho	0.1
Venda	Venda	0.6
Rest of South Africa		86.5

Example 4: Active numbers: Understanding South Africa ...continued

(f) Classification into one of four groups in South Africa

The Population Registration Act of 1950 (repealed in 1991) classified everyone in South Africa into one of four groups:

- **Black:** all the ancestors of these people came from the African continent (all the cards from 1 to 23 inclusive).

- **Coloured:** this group included many different groups including those whose ancestors came from the former Dutch East Indies, Chinese and those of mixed ancestry (cards 24, 25, 26).

- **Indian:** this group included those whose ancestors went to South Africa towards the end of the nineteenth century to work on sugar plantations (card 30).

- **White:** according to the 1950 Act, a white person is someone who 'in appearance obviously is a white person and who is not generally accepted as a coloured person'. For the purpose of this activity the people classified as white are those whose ancestors all originally came from countries in Europe (cards 27, 28, 29).

(g) South Africa – former homelands

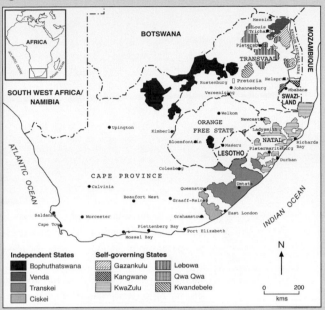

OTHER ACTIVITIES IN BRIEF

Categorising quantitative data

One way of making sense of quantitative data is to sort them in some way, i.e. putting into a rank order or putting into categories and sub-categories. The categorising activity can be done in a variety of ways:

- Students are provided with statistical information on separate cards. The task for the students is to discuss, in groups, different ways of sorting the cards into groups and to discuss ways in which this could be represented graphically (on maps and/or graphs).

- Students are provided with a sheet of statistical data, e.g. related to wards of a city. Students, working in pairs, select a different indicator and sort the data. The students plot the groups on choropleth maps and discuss the patterns.

- Students are provided with information in Microsoft *Excel* format. The use of *Excel* would speed up the process of ranking data and graph production, leaving more time to interpret the data. Where *Excel* is linked to maps through geographical information systems, the sorted data could be mapped easily.

Labelling graphs and maps

This activity is similar to a DARTs diagram completion activity. The purpose of the activity is to help students make sense of what the graph/map shows. Students are provided with a graph or map together with labels of some of the key features. The labels would be related to numerical features of the graphs or maps, e.g. highest figure, lowest figure, range, trends, patterns, etc. Students work in pairs, attempting to match the labels with the graphs/maps, before making their own copy. This activity is suitable for a wide range of graphs and maps, e.g. population pyramids, climate graphs, hydrographs, flow line maps, maps of distributions.

Talking graphs and maps

The activity of talking graphs (or maps) develops students' ability to interpret statistical information presented on graphs or maps. Students work in pairs and then small groups. The procedure is as follows:

1. Students, working in pairs, attempt to identify the main features shown on the graph/map, perhaps with a specific target, e.g. 5 or 10 key features, depending on the graph or map. They list these as bullet points.

2. The pairs join up to form a group of four. They compare the bullet points and try to agree the most important points.

3. In a whole-class activity the teacher collects one point from each group and then adds to the list until all the points have been made. The class tries to agree on a list of key features. Students in pairs use the shared information to try to write a clear paragraph about what the graph shows. Some of these paragraphs are shared with the class.

Debriefing of the activity: What does the graph show? What kinds of things do you look for when you are trying to understand what the graph (or map) is saying? What does the graph not reveal about this topic?

Questionnaire survey

One way of collecting data for both fieldwork and classroom-based geographical enquiry is through a questionnaire survey. Questionnaire surveys can enable students to be actively involved in the complete processes of geographical enquiry and the data-handling cycle of the mathematics curriculum. Examples of the use of questionnaire surveys can be found in Chapter 12 (see pages 157-163).

SUMMARY

Geographical enquiry work at key stage 3 requires considerable mathematical understanding. Most of what is required in geography is also introduced in mathematics lessons, so there is a need for geography teachers to be aware of what students learn in mathematics and when. The processes of enquiry and data handling are as important in mathematics as in geography so there are many opportunities for mathematics and geography departments to liaise and to work towards consistency of practice. Geography teachers can develop activities which encourage students to speculate, to collect numerical data, to represent numbers graphically and to make sense of them through discussion. Although the primary purpose of such activities is to promote the development of geographical enquiry skills and understanding, enquiry work in geography provides a purposeful context in which students can also enhance their understanding of numeracy.

PART TWO
Focused Enquiry

Margaret: What did you do
at school today Sam?
Sam (then eight years old):
What do you want to know?

'We all live in the same world, yet some people seem better able to respond to that world than others, better able to describe the world, to write about it in detail and to re-create it with greater precision. In so doing they enable us to see and understand things better'

(Abbs and Richardson, 1990, p. 44).

INTRODUCTION

Descriptive information about a place or a topic provides a foundation for most geographical enquiry work. This chapter focuses on describing as a worthwhile and challenging focus for enquiry work. It outlines the range of questions that seek descriptive information, identifies some of the skills needed to answer them and presents activities that can foster them.

Of course, enquiries that focus on describing are not sufficient in themselves for examining geographical issues. Most issues in geography are best explored through addressing all the essential enquiry questions of *what, where, why, with what impact* and *what ought* (see 'Schools Council 16-19 Geography Project', pages 20-22). It can, however, be worthwhile to focus, for a lesson or so, on the development of the particular skills involved in descriptive work and to aim to include higher levels of thinking. The questions and activities suggested in this chapter could form just one lesson of a unit of work that would also include lessons focusing on the other essential enquiry questions.

KEY QUESTIONS

Enquiry focusing on describing typically answers questions starting with *what* and *where* for example. These can be expanded to include questions such as:

- What is this place/situation/structure like? When? Where? For whom?
- Where is it?
- How many/few are there?
- Is there a pattern in the distribution?
- In what ways are these places/situations/structures similar and in what ways are they different?
- To what extent is this place more economically developed/less economically developed?

Key questions could also include speculative questions such as:

- What might this place/situation/structure be like?
- Where might this be?

LEARNING OPPORTUNITIES

Most of the thinking skills shown on the framework for geographical enquiry in Figure 3, Chapter 3 (page 44) can be incorporated in enquiry work focused on describing. In particular, such enquiries provide students with opportunities to:

- extend their factual knowledge of places and geographical themes
- develop map reading and map making skills
- describe both orally and in writing
- select
- categorise
- compare
- contrast
- analyse
- interpret
- synthesise
- speculate.

THE CHALLENGE OF DESCRIBING

There is a range of classroom activities through which students could answer key questions beginning with *what* and *where*. These can range from extremely undemanding activities requiring little in the way of thinking to activities which can be very challenging. It all depends on the level of thinking required by the activity. Two frameworks provide some guidance on levels of thinking. The framework in Figure 1 presents a hierarchy of thinking skills related to describing and is based on Bloom's (1956) Taxonomy of Objectives. Barnes (1982) devised the framework shown in Figure 2. Both these hierarchies suggest

Figure 1: Levels of thinking related to describing. Based on: Bloom, 1956.

Levels of thinking	Description	Activities	Typical questions
1. Knowledge	Remembering previously learned information Terminology, specific facts, principles	• Copy • Repeat • Underline • List • Name • Label • Recall • Remember • Find information (with limited need to search)	What is the term used for? What is the capital of? Where is the River Ganges? What is a meander?
2. Comprehension	Understanding or grasping the meaning of something	• Describe in own words • Translate from one form to another (e.g. from a map into words) • Summarise main points • Find an example of • Locate • Select • Illustrate	(From map) Which parts of the country have the most rainfall in winter? (From photograph) What are the main attractions of this place for tourists?
3. Application (Higher order)	Ability to use what has been learned in a new situation	• Apply categories to describe something new • Use information to forecast the future	What is the pattern of weather in Europe? (applying categories used for UK) Given the current rate of growth, what will the world population be in 2050?
4. Analysis (Higher order)	Breaking down into component parts, seeing relationships between parts	• Analyse • Classify • Categorise • Compare • Contrast • Distinguish • Differentiate • Order/arrange	How would you classify …? What are the similarities and differences between …?
5. Synthesis	Creatively applying prior knowledge and skills to produce something new	• Write a report • Design a plan • Hypothesise	How would you redevelop this part of the town?
6. Evaluation	Judge the merit of an idea, or of work against criteria	• Evaluate work • Criticise • Defend work • Justify choices made	

Figure 2: Framework for analysing levels of thinking. Source: Barnes, 1982, p. 196.

Level I

1. Reproducing information from texts with little or no modification
2. Recalling

Level II

3. Measuring and recording numerical data
4. Categorising (identifying and naming)
5. Reading to find specific information
6. Applying procedures (for example, calculation)
7. Describing observations (without interpretation)
8. Translating from one medium to another (for example, map to verbal)

Level III

9. Summarising information
10. Planning (to test hypothesis, etc.)
11. Narrating (with specifications, imaginative projection, etc.)
12. Describing (within a frame of reference, but not drawing general conclusions or explanations)

Level IV

13. Interpreting or hypothesising (placing phenomenon in a theoretical framework)
14. Applying principles (to an unfamiliar type of problem)
15. Problem-finding (formulating issues to investigate)

that describing can involve higher level thinking. It is, of course, possible for students to answer questions starting with *what* and *where* by reproducing information from a textbook, a resource sheet or frequently asked questions (FAQs) list on the internet. This type of activity requires minimal thinking and would be at Bloom's lowest level, curiously termed 'knowledge' and at Level I in Barnes' framework. Students could alternatively answer the same 'What?' and 'Where?' questions by searching data for key points or by using their own words to describe a picture or a distribution on a map. To do this, they have to make some sense of the information rather than regurgitate it as in Level I. There are also activities associated with description at higher levels of thinking: using a framework of categories in relation to new data (e.g. 'To what extent?' activity); doing a piece of extended writing, using information gained from data; devising a plan for something and evaluating a description using criteria.

In some but not all simplifications of Bloom's taxonomy the verb 'describing' is misleadingly included only at the lowest levels. There are different levels of describing and enquiry work can be planned to make them challenging.

ACTIVITIES

The activities suggested below are devised to get students involved in describing for themselves. They all involve the study of data.

- Chinese whispers
- Ready, steady, remember
- What might it be like?
- What and where?
- Comparisons
- To what extent?
- Identifying links
- Layers of meaning
- Internet enquiry

The first three activities encourage close observation of visual data.

CHINESE WHISPERS

This starter activity encourages students to look at visual information and to remember it as a picture in their minds so that they can describe it to students who have not seen the information. The procedure is set out in Example 1. The picture can be a very large picture which all can see, or a projected image from a slide or a computer. The activity enables students to become more aware of what needs to be in a description to portray an image accurately. The listeners, in building up an image in their minds, rely not only on what is said, but also previous knowledge and experience. The words they hear conjure up images from their existing knowledge and sometimes these are very different from what has been shown. They are interpreting what they hear in terms of their existing schema or ways of thinking about the world. This point can be brought out in the debriefing. Those who have seen the picture tend to be totally involved in the activity. They retain the image in their mind's eye and will be aware of what is missing from the new description or of what has been changed. Mentally, they are comparing a description with an image and noting the discrepancies.

Example 1: Chinese whispers.

Procedure

- Five students are sent outside the classroom or to a place where they cannot see and hear what the rest of the class will see and hear.

- The rest of the class is shown a slide or a picture. It is best if they are not told where the picture was taken. There could be brief discussion of the main features of the picture. The picture/slide is removed and those who have seen it remember it in their 'mind's eye'. A decision is made on who will describe the picture to the first listener.

- One student, the first listener, comes back into the classroom.

- One or more people in the class describe what they have seen, but without giving specific information about where the place might be. The first listener cannot ask questions, but relies on what is said. He/she builds up a picture in his or her 'mind's eye'.

- A second student comes into the classroom and the first listener describes what he/she has heard about the picture, i.e. the image conveyed by the description. No additional help is given by those who have seen the picture. A third student comes into the classroom and is told by the second listener what he/she has heard indirectly about the picture.

- This process continues until all five students outside the room have returned.

- The last student is asked to describe what he/she expects to see when the picture is shown again (or possibly to draw a rough sketch).

- The picture is shown again.

- Some possible debriefing prompts:

  1. (to the listeners) What were the most useful descriptions? What words and phrases were useful? What picture did you build in your mind? Why? Where did these images come from? Were you thinking of particular places?

  2. (to the class as a whole) What kinds of things were forgotten and what were remembered? Can you explain this? What kinds of things were changed? Why? How could you have improved your initial description? Where do you think this place is? Why? How reliable is second/third hand information? Can you think of any time when you heard about a place and built up a picture and then the place was totally different?

This is a motivating activity that can be repeated with a different picture. With experience, students will get better at putting pictures into words. This starter activity can be developed into a whole-lesson enquiry, by following it with other activities, e.g. one in which students write descriptive letters or postcards from particular places.

READY, STEADY, REMEMBER

In this starter activity, students are encouraged to look closely at visual information because they know that they have a limited time to study it. This increases concentration and helps students develop the skills of observation and putting images into words. The kind of data that would be suitable would be a picture, graph, diagram, very short video extract (no more than 2 minutes in length), or a map. A possible procedure is shown in Example 2.

Example 2: Ready, steady, remember.

Procedure

- Students are introduced to the topic.

- They are told that are going to have a very limited time to study some data and are given a triple challenge: How observant can they be? How much can they remember? How good can they be at putting into words what they have seen?

- They are shown a picture, graph or diagram for a period of 15-30 seconds, or a very short extract from a video (no more than 2 minutes). The students observe but do not write anything down.

- They discuss in pairs what they can remember seeing and write down key pieces of information.

- They share the information as a class, to see how much they have been able to observe and remember in such a short time. Are they sure about what they remember? (No additional information is provided by the teacher at this stage.)

- They are asked if they want to see the image/video again (the answer is always yes). They observe again.

- Debriefing prompts: What kinds of things did you remember? What kinds of things did you forget? Did other people remember this? Which pieces of information do they think are important geographically? Which pieces of information are not significant geographically (e.g. the colour of a car)? Why? What categories of information are there?

- Students work in pairs writing descriptive sentences about the image. If there are several different categories of information, different pairs can work on different aspects of the image.

- Students share their descriptive sentences.

This starter activity can be developed by following it with the study of more images or the rest of the video. This can be followed with support for producing high quality extended descriptive writing. In Example 1a, Chapter 10 (page 130), Ready, steady, remember is followed by a DARTs activity.

WHAT MIGHT IT BE LIKE?

In this activity students are provided with an initial stimulus and a context, and they are then invited to speculate on what somewhere or something might be like before they find out what it is like. An example of the use of this activity, in relation to a descriptive enquiry focused on Antarctica, can be seen in Example 3, Chapter 7 (page 86).

WHAT AND WHERE?

In this activity students are provided with a context in which they need to study where something is located. The activity makes them think about location by requiring them to transform the information from one form to another:

- either they are provided with a distribution on a map, from which they write a description,

- or they are provided with information in a table or in text and they transfer this information to a map.

In order to carry out the activity they need to make sense of the information. Example 3 illustrates an enquiry into the distribution of UK weather, where the initial information is on a video recording. The enquiry helps students develop the skills of using both symbols and words to describe distributions. The regrouping of students into new groups and the sharing of data collected in the first group encourages co-operation between groups and communication skills.

Example 4 illustrates an enquiry into the distribution of earthquakes where the initial information is obtained from the internet. This enquiry gives students the opportunity to search for up-to-date information and to develop the skill of locating places using atlases and figures for latitude and longitude.

COMPARISONS

Making comparisons is a feature of everyday life. Students routinely compare people, foods, leisure activities, schools they have been to, teachers in schools; it is a capacity that they already have. Yet many students find it difficult to compare and contrast in geography. This is partly because comparisons in geography are often presented as an academic exercise which might be of little interest or relevance to some students; it is just something they are asked to do. It is also partly because in everyday life they talk

Example 3: Describing the weather.

In this enquiry students answer the question: What is the weather like in different places? The data is provided on a video recording of a UK weather forecast. Outline maps, possibly drawn from a projection onto large sheets of paper, are also required.

Procedure

Starter

- Question and answer: Who needs to know about what the weather will be like? Why would they need to know? How would they find out?
- Challenge: How good can everyone be at observing and remembering information from a weather forecast? How many times will the forecast need to be played before the class as a whole has extracted all the information?
- Students watch the video once without taking notes.
- Brief interim discussion: What different aspects of weather were described (e.g. temperature, cloud cover)? What units of measurement are used for each aspect? What places did they mention? Why is it difficult to say what the weather is like in one place? (Usually the weather will change over a period of time.)

Group work activity: Collecting data

Preparation for focused, collaborative collection of data. Divide the class into 5 or 6 groups. Each group is instructed to watch the weather for a specified region, e.g. Wales (point out all the regions on a wall map). All groups are instructed to collect data for specific times, e.g. one day and the following night.

- Students watch the video focusing on their particular region and times without taking notes.
- In groups they write down what they can remember of the weather and the places mentioned.
- Students watch the video again, concentrating on watching.
- They add to their notes. If necessary, they watch again.
- The group checks that each person has all the information from that region.

Group work activity: Recording data

- Students regroup with one person from each region in each new group.
- Each person is responsible for telling other members in the group what the weather was like in their region.
- Students use symbols to record the weather information onto an outline map.
- They name some of the places referred to in the weather forecast.

Plenary activity: Presenting the data and debriefing

- The groups present their maps to the whole class, describing what the symbols show and commenting on the distribution, using appropriate terminology and place names.
- Debriefing prompts: What did you find hard about watching? Why? Which were the easiest things to see? How would you describe the weather in, for example, Scotland, for this day? Did any parts of the country have different weather? Why do you think there are differences in the weather in different parts of the country? (This encourages speculative thinking, to be followed up in later enquiry work.) What changes were there in the weather over time in your area? Why do you think the weather changes during the course of the day (encouraging speculative thinking)? How accurate do you think the information is? Why or why not?

Example 4: Where in the world have there been earthquakes?

Key questions

- Where in the world have the most recent earthquakes occurred?
- Is there a pattern in their distribution?
- Does this pattern match the pattern of earthquake distribution over the years?

Resources

Information on latest earthquakes (preferably on-line, otherwise printed out) from, for example:

- US Geological Survey, National Earthquake Information Center – http://neic.usgs.gov/neic/bulletin
- World-wide earthquake locator – http://www.geo.ed.ac.uk/quakes/quakes.html

Starter activity: Intelligent guesswork

Questions to encourage speculation: How often do you think earthquakes happen in the world? How many do you think there have been so far today? Are they more likely to happen in some parts of the world than others? In which parts of the world do you think they are likely to happen? In which parts of the world do you think they are unlikely to happen? (Make notes of suggestions on board for later use). On what scale are earthquakes measured? What do you think a large earthquake would measure?

Activity: Searching for and collecting data

- Students search the internet to find the most recent earthquakes.

- On a prepared sheet they make notes of where each earthquake occurred and its magnitude.

Interim plenary discussion (or discussion with individuals as they carry on with both activities)

Questions to guide the discussion: Did you find out how often earthquakes happen? Where in the world have the most recent earthquakes been? Which ones would you classify as large? Which would you classify as minor earthquakes?

Activity: Recording the data

Students plot the data on a world map either from latitude and longitude (NEIC site) or from place names or from maps on the internet.

Plenary discussion and summary

- Questions to guide discussion: What do you notice about the distribution of recent earthquakes? Is there any pattern? Are there any continents or large regions of the world without any earthquakes at all?
- (Using an atlas map of distribution.) How does your map compare with the atlas map? In what ways is the pattern the same? In what ways is it different?
- In pairs: write down four key points about the distribution of earthquakes.
- Share this information and work out collaboratively some general points that everyone agrees on.
- Students write down the key points or annotate the world map with them.

comparisons rather than write comparisons. In developing the descriptive skills of comparing and contrasting at key stage 3, it is worth thinking carefully about the procedure for developing the enquiry. A possible procedure is shown in Example 5.

There are innumerable opportunities for comparative enquiries at key stage 3. Students may find it easier to develop comparative thinking skills if the activity they are asked to do relates to comparisons people make in the world outside the classroom, rather than treating it as a purely academic exercise. It would be worthwhile devising a context in which there is a need to compare and contrast, for example:

- destinations for a day out or for a holiday
- where to go shopping
- which area a family might prefer to live in when moving house.

Example 5: Enquiries that compare and contrast.

(a) Procedure

Key questions

- In what ways are these places/situations/events similar?
- In what ways are they different?

Resources

- Data which provides comparable data on two places or situations or events
- Note-taking frame (see Examples 5b and 5c)
- A purposeful imagined context

Starter

- Ask students in pairs to compare two things they all have experience of, e.g. primary school and secondary school.
- Collect ideas and put them on board in a framework similar to the note-taking frame to be used (Venn diagram or T diagram).
- Use the terms compare, similarities, contrast, differences.
- Introduce students to:
  1. the context in which they need to compare and contrast data for a particular purpose
  2. the data they are going to use to compare and contrast
  3. note-taking frame

Activity: Collecting data

- Students work in pairs on the data.
- They record similarities and difference on a note-taking frame.

Plenary: Preparation for writing a report

Feedback from data collection (reminds them of the context in which the activity was set).

Depending on context, debriefing questions could include: What similarities did you find? (List on board.) How might these be categorised? Which do you think are the most significant similarities? Who would they be important to? What differences did you find? How might these be categorised? Which do you think are the most significant differences? Who might these differences be important to? How did you set about looking for similarities and differences in the data? Could you improve this strategy? What are the most important points you have learnt? How could these be organised into a report? What might each paragraph be about? How could each paragraph start? What link words can you think of?

Activity: Writing a report

- Students write a report about similarities and differences (preferably aimed at an imagined audience that needs to know).
- Reports could be shared in a display or in a later lesson.

(b) Note-taking frame for comparison: Venn diagram

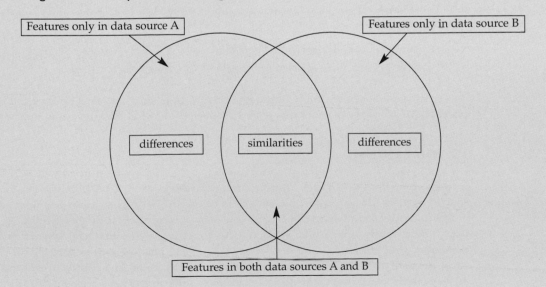

Example 5: Procedure enquiries that compare and contrast ... continued

(c) T diagram note-taking frame for comparing and contrasting

Similarities between _____ and _____

1

2

3

4

Differences between _____ and _____

_____	_____
1	1
2	2
3	3
4	4
5	5
6	6

Another possibility is to compare what is being studied with the students' own experience. For example, a study of photographs of a less economically developed country, or indeed of any other place, can be framed by the questions:

- In what ways is this place similar to England/the area in which you live?
- In what ways is this place different from England/the area in which you live?

This activity encourages students to relate new knowledge to what they already know and to relate to places which at first glance might appear very different, but in which there are many similarities.

TO WHAT EXTENT?

In learning to describe, students learn to categorise. Often, however, categories in geography over-simplify matters. This is particularly true of the categories 'less economically developed' and 'more economically developed'. It may be relatively easy to categorise countries from generalised statistical data. Most countries in the world, however, have a great variety within them; some areas are highly developed while other areas are less developed. Labelling countries such as South Korea, Brazil and China as LEDC, can give students a misleading impression; parts of Seoul, Rio de Janeiro and Shanghai are more developed economically than anything most students will have encountered. Most countries that would be classified as 'more economically developed' have areas which are economically less developed. Students can get a more realistic picture of the world if they can be made aware that countries are more or less developed only to a certain extent.

In studying LEDCs and MEDCs, it is worth extending descriptive enquiries to ask the question: 'To what extent is this place developed?'. Example 6 illustrates how this activity can be developed in the classroom.

Example 6: The To what extent? activity. Source: Caroline Young.

(a) Procedure

Key question

To what extent is Kenya/Pakistan (or other LEDC) economically developed?

Resources

Visual data provided by a photopack or a video.

Starter

- Ask students to write down what kinds of things they think of when they hear the phrase 'less economically developed country' (they could do this in pairs).
- Ask students to write down what kinds of things they think of when they hear the phrase 'more economically developed country'.
- Students feed back their ideas to the whole class and the teacher builds up two spider diagrams on the board, one of LEDCs and one of MEDCs.
- Students are asked whether they agree with what is in each. If necessary modifications are made.

These diagrams represent students' existing thinking, and become a model to apply to the next activity.

Activity: Collecting data from video or from photographs

- Either students study photographs and jot down evidence on a note-taking frame (Example 6b was devised for use with a photopack of Kenya),
- or students watch a short section of video, without taking notes, looking for visual evidence and listening for verbal evidence. The video is stopped and they write notes in a note-taking frame. The next section of video is shown, etc.

Interim plenary discussion

- What evidence is there that this country is 'less economically developed'? Why does this make it 'less economically developed'?

- What evidence is there that this country is 'more economically developed'? What does this mean? Was it in particular places? Did you see anything that surprised you? Why did this surprise you? In what ways is this country similar to the UK?
- On balance what conclusions do you draw from the evidence?
- Do you think this evidence is likely to be fair? Would it be possible to produce evidence for this country which would show it as an MEDC? What would you select to show it as an MEDC? What would you select to show it as an LEDC (this could be followed up with an analysis of photographs of this country in the textbook being used)?

Making sense of data: Extended writing

- Students are prepared to write a report, 'To what extent is Kenya/Pakistan (whichever country is being studied) economically developed?'
- Plenary discussion: How could this report be structured? What would need to be in the introductory paragraph? What would each paragraph be about? How could each paragraph start? What examples might you choose to illustrate what you are saying? What would be in the final paragraph? (Ideas can be shared and put on the board.)
- Students write the report, if necessary supported by a writing frame (Example 6c).

Reflecting on learning

- Sharing written reports. Students, in pairs, compare their reports. They note down one point they both made and one difference between the reports.
- Plenary discussion: What similarities were there between the reports? What differences were there? What do you think now? To what extent is the area you have studied economically developed? Has this enquiry changed your views about this place?

Example 6: The *To what extent?* activity. Source: Caroline Young ... continued

(b) Writing frame

How developed is ....?

There is some evidence that .... is economically developed:

1 ....

2 ....

3 ....

4 ....

On the other hand, there is also evidence that .... is less economically developed:

1 ....

2 ....

3 ....

4 ....

In conclusion I think that .... is more developed/less developed (cross out one of these) because:

1 ....

2 ....

(c) Task sheet

Development

To what extent is Kenya developed?

Task

1. Look at the photographs of Kenya.

2. Look for evidence of development. List the evidence in column A, noting down the number of the photograph providing this evidence.

3. Look for evidence that Kenya is 'less developed'. List the evidence in column B, noting down the number of the photograph providing this evidence.

A. Evidence of development in Kenya	B. Evidence that Kenya is less developed

Homework: To what extent is Kenya developed?

Write three paragraphs. You could use the sentences and phrases below to start your paragraphs.

In some ways Kenya is developed. In the photographs I saw evidence of ...

In other ways Kenya is less developed. In the photographs I saw evidence of ...

My overall impression of Kenya is ..., or On balance I think that ...

IDENTIFYING LINKS

In this activity students answer the question, 'What links are there with other places?'. This question can be applied to their local area or town, this country or themselves, or to some aspect of their own lives such as clothing or food or leisure. The data might be their existing knowledge and experience or it might be based on some individual research (into food labels, clothes labels, places in the local area). Students identify the places with which there are links. In order to make sense of the links, they could categorise the information, map it and look for distribution patterns. The answer to the question will be descriptive, but the enquiry work could be extended to seek explanations. Example 7 illustrates how this activity was used as an introductory activity to an enquiry into differences within the European Union.

Example 7: Identifying links with countries in Europe. Source: Carol Webb.

(a) Procedure

Key questions

- In what ways is the UK connected to other countries in Europe?
- Which of these links are with countries in the EU?
- What is interdependence?

Resources

- Worksheet: How are we connected to Europe? (Example 7b)
- Props: for example, pictures of Nokia mobile phone; footballer playing in the Premiership; fashion article, holiday brochure
- Data: students' existing knowledge

Starter

- Give out worksheet.
- Students underline all the countries they have heard of (ensures reading and reinforces existing knowledge).
- Puzzle: Why are some countries in the circle and some outside the circle?
- Challenge: Will you be able to think of a link between each country and the UK? What kinds of links are already identified? What other kinds of links do we have with these countries (e.g. music, food, sports, clothes, personal links – use props)? Will you be able to think of links in each of these categories?

Activity: Identifying links

- Students discuss the work in pairs or small groups.
- Students write the links they have thought of on the worksheet as shown in the completed worksheet (Example 7c).

- Extension activities for those who have finished or who have exhausted their ideas:
  1. They categorise the links by colour coding them and devising a key.
  2. They make links between other European countries, e.g. between France and Germany.

Plenary discussion

- Sharing what students have marked on their diagram (correcting if necessary).
- One link for each EU country (students can add to their diagrams).
- One link in each of the following categories: sport; music; clothes; cars; food; travel (students can add to their diagrams).
- Debriefing: Are there any countries that were difficult to think of links for? Why? Are there any where no one has found links? (Go through the more difficult ones.) Which countries were easiest to think of links for? How does the EU strengthen links (e.g. passports, work, laws, agriculture)? In what ways would our lives be different if we did not have these links? What would you miss most? How dependent are we on these countries?

Activity: Writing

- Prepare for writing summary notes and sentences about personal feelings. What are the main things we have learnt about links with European countries?
- (List on board.) How can we organise all these points into notes? How important are these links to you? Which links would you miss most?
- Students write notes and summary sentences.

(b) The 'How are we connected to Europe?' worksheet

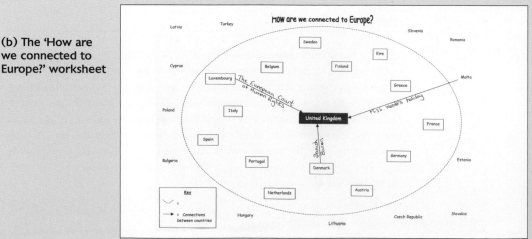

Example 7: Identifying links with countries in Europe. ... continued

(c) Sophie and Sam's completed worksheet

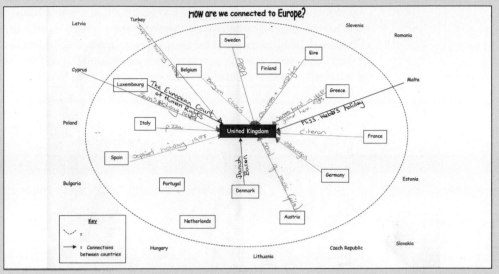

Example 8: Layers of meaning framework.

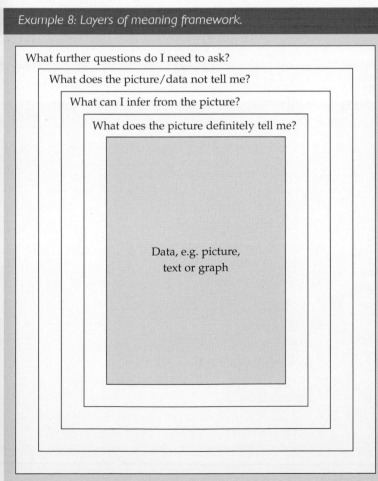

LAYERS OF MEANING

This activity has been widely used in history lessons to study both verbal and visual source materials (Cooper, 1992; Riley, 1997). The framework of four questions (Example 8) was originally designed for use in primary schools to encourage children to consider the usefulness of historical sources on a variety of levels. The central question demands the detection of what is certain. The next question challenges students to distinguish between what the data definitely say and what might be inferred. Riley (1997) refers to these kinds of diagram as 'layers of inference'. The outer questions make students aware that the data presented are selected from a particular viewpoint, and invite students to reflect on this by adding their own questions. The activity encourages students to delve into different layers of meaning, each with different degrees of certainty. The questions can be set in worksheet form on a sheet of A4 paper, or an A3 sheet with space to write all around the data. This activity can be used with text, photographs or graphs.

INTERNET ENQUIRIES

The Internet is an excellent source of descriptive information both verbal and visual, but collecting this information does not turn the activity into geographical enquiry. Internet enquiries, like other enquiry work, has four essential aspects. First, there has to be a need to know, a key question to be answered, and a stimulation of curiosity about the issue being investigated. Students can then search the internet for relevant information, possibly supported by a note-taking frame. Then some sense needs to be made of the information collected. Finally, it is useful to reflect on the information that has been collected and its reliability as evidence to answer the question.

Example 9 shows a note-taking frame for a preliminary internet enquiry about Swanage in Dorset to answer the question: 'What is the area around Swanage like?'

Example 9: A note-taking frame for internet searching.

Where is Swanage?	**A brief history of Swanage**	**What activities are there in the Swanage area?**
What can you find out about flooding in Swanage?	**What is special about the Swanage coastline?**	**An image of Swanage or the surrounding area**

Websites: www.virtual-swanage.co.uk or www.swanage.gov.uk

Example 10: Framework for selecting descriptive information on Antarctica from a range of resources.

Why is Antarctica so special?

Over the next two lessons you will be carrying out a research project on Antarctica. You will be working in a group of four people and will produce a poster showing information about Antarctica and why it is so special. Your poster must include the following:

- Introduction to Antarctica (e.g. size, climate, landscape)
- A map showing the location of Antarctica
- Information on environmental, social, political and economic aspects of the continent. The key questions in the table below might be useful.

Environment	Social
What is the climate like? What wildlife is there in Antarctica? Why are these species so special? What is the landscape like?	Who lives on Antarctica? Where do these people live? Does Antarctica have tourists? What are the environmental issues?
Political	**Economic**
Who does Antarctica belong to? Who manages Antarctica?	What resources are there? Should we be allowed to exploit Antarctica?

You should use a variety of ways of displaying the information (e.g. writing, pictures, maps, graphs, tables) and make maximum use of the resources available to you (don't just use books or the internet).

The resources available are:

- the internet
- library books
- CD-Roms (World Book, Britannica, Encarta, etc.)
- Videos

Useful websites

- http://www.lonelyplanet.com/destinations/ antarctica/antarctica/
- http://www.cia.gov/cia/publications/factbook/ geos/ay.html
- http://www.terraquest.com/va/science/about.html
- http://www.antarcticanz.govt.nz/Pages/ InfoEducation/Infosheets.msa
- http://www.asoc.org/

The purpose of this enquiry was to collect some background information prior to a decision-making activity on possible flood prevention schemes in Swanage. The note-taking frame encouraged students to search for particular information and to be concise, and it gave them the opportunity to choose a photograph from the many on the websites for themselves. This example of a note-taking frame for collecting descriptive information from the internet or from other sources can be adapted and applied to innumerable topics investigated in geography.

Example 10 shows the use of a framework to encourage the selection of descriptive information on four different aspects of Antarctica: environment; social; political; economic. Students are expected to select from data on selected websites, in the library, on CD-Roms and on videos.

LITERACY SUPPORT

To write well descriptively is by no means a lower order thinking skill. The best descriptive writing in novels, travel literature, newspapers and in radio and television reports demonstrates the highest order of thinking. In the classroom, describing can be a challenging activity. The skills of describing can best be developed in purposeful contexts in an environment in which good descriptive speaking and writing is enjoyed and valued. The main purpose of descriptive writing is not to enable students to complete exercises correctly but to enable them to 'see and understand things better' (Abbs and Richardson, 1990, p. 44). There needs to be a focus on meaning and purpose as well as on form.

Students' literacy skills can be developed by:

- providing opportunities for classroom talk in which describing is important
- providing students with opportunities to listen to, to see and to read descriptive writing of all kinds
- encouraging students to read travel literature and to search for information in encyclopaedias and on the internet
- incorporating in enquiry work a need to report the findings orally or in writing
- providing opportunities for all kinds of descriptive written work which could include the writing of reports, postcards and letters, brochures, and newspaper accounts
- encouraging the use of appropriate geographical terminology
- encouraging the use of appropriate adjectives
- encouraging students to use geographical vocabulary related to location, e.g. in the northern part of …
- encouraging students to compile their own glossaries or personal geographical dictionaries
- providing for students who need it additional support in the form of lists of vocabulary, connective words and sentence starters (Figure 3).

Figure 3: Literacy support for enquiry work with a focus on describing.

Students who need additional support for written work could be provided with the following:

1. Geographical terminology related to what is being studied, including categories such as economic, social, environmental
2. Adjectives
3. General vocabulary which supports descriptive writing, e.g. for instance, such as, for example, yet, however.
4. A few phrases and sentence starters can be devised for a particular enquiry, for example:

- Before I studied this enquiry, I thought …
- Now I have found out that …
- The most interesting thing I have found out is …
- The thing that surprised me most was …
- In some ways … and … are similar …
- They both …
- In other ways they are different from each other …
- In contrast …
- On the other hand …
- Further north …
- In the east of the country …
- In particular …
- Whereas …
- Unlike …
- Apart from …
- In some ways …
- In other ways …
- On balance I think that …

'What causes something to
happen has nothing to do
with the number of times it
has happened or been observed
to happen and hence with
whether it constitutes
a regularity'

(Sayer, 1985, p. 162).

INTRODUCTION

Explanation is at the heart of geography. The subject of geography developed because of curiosity of why the world is as it is and the need to search for explanations. Most enquiry work that involves explanation will also involve some description of what is being explained. However, the words 'describe' and 'explain' are so linked in examination questions that it is worth separating them for some enquiry work in order to focus on what it means to explain.

KEY QUESTIONS

Questions which seek explanations in geography are varied. They include questions about physical processes and also questions about human intent, motivation and action. Questions seeking explanations and thinking about processes include:

- Why is it like this? How did it get to be like this?
- What effects are there? What are the consequences of these effects? What are the implications of these effects?
- What factors influence this? Which factors were most important?
- What explanations are there for this situation?
- How can we interpret this situation?
- Who had power to influence the situation?

Key questions could also include speculative questions such as:

- What explanations could be suggested for this situation?

LEARNING OPPORTUNITIES

Enquiries which focus on explaining provide students with opportunities to:

- develop understanding of processes in human, physical and environmental geography
- become aware of some explanatory theories
- become aware of the complexity of explanation in geography
- recognise links between different sets of geographical information
- explain both orally and in writing
- compare and contrast
- correlate
- reason
- speculate.

THE CHALLENGE OF EXPLAINING

Explanation can be challenging. The degree of challenge depends on what kinds of thinking students are expected to engage in when they explain. The prefixes *how* and *why* do not in themselves make questions challenging. If reasons are listed in books or on the internet and the only expectation is that students copy them down, then minimal thinking is required.

Figure 1, based on Bloom's taxonomy of objectives, provides a possible framework for considering levels of thinking related to explanation. In order to think at higher levels, students need to consider existing explanations and understand them. They need to examine data as evidence for explanations and theories. They need to become curious about the world and speculate on why things are as they are.

They need, as far as is possible, to get into the minds of those who searched for explanations in the first place. Students need to use their minds to reason and not simply to learn lists of reasons.

Explanation in geography is a complex issue. Some explanations in geography are scientific. Geographers looking for scientific explanations are seeking to produce general laws of how things work and to be able to use these laws to make predictions. Other explanations in geography are interpretations of complex unique situations which are not inevitable and which could not be predicted. They are situations in which human intentions, values, decisions and power relations have to be taken into account in explaining the situation. There can be different and conflicting theories to explain the same situation. We therefore have to plan enquiry work focusing on explanation with great care, so that students do not understand the world in a deterministic way.

Figure 1: Levels of thinking related to explaining. Based on Bloom's hierarchy of six levels.

Levels of thinking	Description	Verbs	Exemplar activities
1. Knowledge	Recognising or remembering previously learned information.	Recall Remember Copy Repeat Underline List Find information	Finding, without needing to think, a reason listed in a textbook, and copying this Cutting and pasting from a website an answer to a FAQ 'why' question
2. Comprehension	Understand or grasp the meaning of something	Explain Translate from one form to another Summarise Make connections between two sets of data Interpret	Explain in own words 'Translate' a text into a flow line diagram Suggest locational reasons from data on sets of maps
3. Application (Higher order)	Ability to use what has been learned in a new situation.	Apply categories	Apply locational factors to new situation
4. Analysis (Higher order)	Breaking down into component parts, seeing relationships between parts patterns	Analyse Distinguish Classify Differentiate Categorise Order/arrange Compare Correlate Contrast	Identify causes and effects and long-term implications in an account of flooding Compare the effects of earthquakes in different places
5. Synthesis	Creatively applying prior knowledge and skills to produce something new	Speculate Write a report Produce	Speculate on causes from evidence Write an illustrated report for television news explaining a news event, e.g. earthquake
6. Evaluation	Judge the merit of an idea, or of work against criteria	Evaluate work Discuss evidence for theories Criticise Defend work Justify choices made	Students evaluate each others' work

The issue is further complicated by the fact that students learn about explanation in other subjects in school, but it can mean different things. In science students may learn about cause and effect as a scientific process, which in some cases is reversible. In history students may consider different explanations for a situation after interpreting data as evidence. In geography when we talk about explanation we sometimes mean scientific explanations and we sometimes mean interpretations of unique situations. It does not help students to understand the world they live in to make them think that there are rules of cause and effect that can explain everything in geography. We need to help students to be aware that there are different interpretations of the world and that different people might see and explain things differently.

ACTIVITIES

The following activities all involve students in using data as evidence and in thinking about explanations:

- sequencing
- categorising
- diamond-ranking factors
- comparing and contrasting
- clues.

SEQUENCING

For many topics in geography, explanation is related to change over a period of time. Continuous change is often broken down into phases dominated by particular processes. A useful activity is to get students to think carefully about sequence and about the processes involved is a reconstruction DART (see Chapter 5, pages 55-57). Students are provided with information about each stage in the sequence, preferably including both text and visual data (maps, photographs or diagrams). The challenge for the students is to think about what processes are taking place at each stage, to match the text with the visual information and to put the data into an appropriate sequence. Although this seems a straightforward task, in practice most students have to study the data carefully and think about processes. Example 1 shows an enquiry focused on the question: How was the African Rift Valley formed? This enquiry, although planned with this key question in mind, was related to the students' own questions generated in the previous lesson in response to a puzzling diagram. This example can easily be adapted and modified for other enquiry work attempting to answer the question, 'How was this formed?'

Example 1: Sequencing using DARTS. Source: Rachel Atherton.

(a) Procedure

Key question

How was the East African Rift Valley formed?

Resources

- Worksheet and statements for DART activity (Example 1b)
- Diagrams for DART activity (Example 1b)
- Over-head transparency (OHT) of questions generated by students in the previous lesson (see Example 1d)

Lesson 1: Ready, steady, remember generating questions

Students work in small groups

- Each group sends one person to the front to look at a diagram of the East African Rift Valley for 30 seconds.
- The person has 1 minute to draw what he/she remembers.
- A second person from each group goes to the front to look at the picture for 30 seconds.
- The second person makes his/her own drawing and labels it.
- The next person from each group goes to the front and then draws the picture, etc.
- Students then look at drawings and think of what puzzles them about what they have drawn.
- Each group thinks of five questions.

Example 1: Sequencing using DARTS. Source: Rachel Atherton ... continued

Lesson 2: Formation of the Rift Valley
Starter

- Students' questions from the previous lesson are shown on OHTs (Example 1d).
- Pairs of students decide which are the best two questions to ask in order to find out more about how the East African Rift Valley was formed.
- Whole class discussion: Which questions did you decide were best? What makes a good question?

Activity: Reconstruction DART

- Students match the statements to the diagrams.
- Students draw a picture to represent the first stage in the sequence (Example 1c).

Plenary

- Questions on OHTs shown again.
- Which questions have been answered by the DART?
- Were there any questions not answered? (Try to answer.)
- Are there any things you do not fully understand? (Try to answer.)
- Which of the questions you devised do you now think were good questions?
- Why?

Homework

- Choose 10 questions and prepare a FAQ section for a website or textbook for students your own age.
- Answer question: What makes a good question? (Example 1e)

(b) Worksheet

How was it formed?

A

The Rift Valley is one of the wonders of the world, stretching from the Middle East, down through Africa, reaching as far as Mozambique (3500km). The staggering view, as you approach from Nairobi, Kenya is quite unbelievable. The ground suddenly disappears from under you to show the huge expanse of the great rift, stretching for thousands of miles in either direction. Whilst this stunning introduction to the Rift Valley is amazing in itself, actually descending and exploring the Lakes area of the Rift in Kenya is a 'not to be missed' opportunity

This passage (A) was taken from an internet source and gives some idea of what to expect if you visited the East African Rift Valley ... but how was it formed?

Kenya is located near to the edge of the African crustal plate (B). The Earth's crust is split into many of these huge plates which 'float' on the underlying molten magma. The collision of these plates often causes the surface of the Earth to buckle into folds and create a fracture or fault. This set in motion a series of events which led to the formation of the Rift Valley (C).

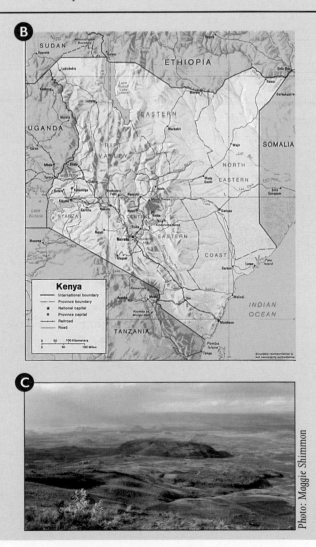

Photo: Maggie Shimmon

Example 1: Sequencing using DARTS. Source: Rachel Atherton ... continued

(c) Richard's explanation of the formation of the Rift Valley

Read the following statements and match them with the diagrams.

- Molten rock rises up through the cracks to form volcanoes along the edges of the Rift Valley.
- Two of the biggest faults were parallel to one another. The land between them collapsed.
- Around 70 million years ago the land in this area was fairly flat.
- Water draining down into the Rift Valley became trapped and formed the lakes.
- The faults or cracks were caused by two plates colliding.

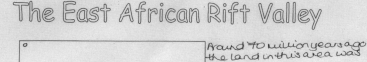

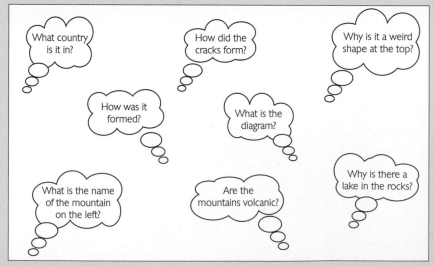

The East African Rift Valley

Around 70 million years ago the land in this area was fairly flat.

A series of collisions with neighbouring plates created a system of faults or cracks in the African plate.

Two of the largest faults were parallel to one another. The land between them collapsed as sections of the crust began to move apart.

Magma from below the surface forced its way up through the cracks to form lines of volcanoes along the edges of the rift valley. Water draining down into the rift valley became trapped and formed a series of lakes.

(d) Questions generated by students

What country is it in?

How did the cracks form?

Why is it a weird shape at the top?

How was it formed?

What is the diagram?

What is the name of the mountain on the left?

Are the mountains volcanic?

Why is there a lake in the rocks?

(e) What makes a good question? – extracts from students' homework

Jedder:

'I'm not quite sure what makes a good question. Though I think if it can answer more than one other question it is good. So if it tells you more than meets the eye then that's great. If the answer goes into lots of detail that's good.'

Tammie:

'In my opinion a good question is a sort of question which gives the answer to a number of other questions straight away and may also give you extra information. Example: Are the mountains volcanic? Quite good question. Example: How was it formed? Good question, because there is more information from good questions than quite good questions.'

CATEGORISING

This activity encourages students to think about causes and effects by asking them to categorise information presented on cards so that they can be sorted and rearranged easily. The cards could be based on information in a textbook or on the commentary of a video. The text is broken into sentences, in a similar way to reconstruction DARTs (see Chapter 5, page 53). Unlike reconstruction DARTs, the task is not to reassemble the text but to categorise it in some way to answer the key enquiry questions. Students can devise their own categories or can be presented with categories such as cause and effect. These categories can be subdivided into further categories, e.g. long-term causes/short-term causes; physical causes/human causes; short-term effects/long-term effects; economic/social/environmental effects. The activity should be devised so that students need to think carefully about explanations. The activity is similar to mysteries (Leat, 1998; Leat and Nichols, 1999) in that the students attempt to answer questions through the analysis of items of information. The activity is different in that the task and the data have not gone through the creative transformation necessary to produce a good mystery. Example 2 illustrates how a categorising activity was used with one year 8 mixed ability class. The text was based on teachers' notes made from the commentary of a television programme on Mount St Helens. The lesson was one of three devised by University of Sheffield PGCE students who were studying the value of different ways of teaching the same content to three different year 8 classes. One class studied a text and answered comprehension questions. One class watched a video and completed a note-taking frame. The third class used the categorising activity in this example. The students categorised pieces of text based entirely on the verbal information in the video. Of the three lessons, the students were most involved in and learnt most in the third lesson. The categorising task made the students think more about the enquiry questions for themselves. In a non-experimental context, the video could be used to provide additional visual data. This could easily be adapted for other explanatory commentaries or texts.

Example 2: Categorising. Source: Anna Mercy.

(a) Procedure
Key questions

- What happened when Mount St Helens erupted?
- What were the effects?

Resources

26 statements cut up and put into envelopes, with one envelope for each group of four students (Example 2b).

Starter: Eliciting existing knowledge

(Answering questions, first in pairs and then to the whole class.)

- What happens when a volcano erupts? Think of four key points.
- What effects can volcanic eruptions have?

(Note on board.)

Challenge: You are going to have data on a particular volcanic eruption. Study the data and make as much sense of it as you can so that you can tell the rest of the class the story of this eruption.

Activity: Categorising cards

- Each group (of four students) is given a set of cards to sort.
- The teacher observes and makes a mental note of how they are sorting the cards and what remarks are made.
- The teacher supports groups through dialogue (e.g. Tell me what you have found out so far? How could these cards be grouped? Why have you put that group of statements together? How does that help sort out the story in your mind?)
- The teacher suggests categories (e.g. cause, short-term effects, long term-effects) to any groups who are struggling to make sense.
- Each group, having categorised the cards, sticks them onto a sheet.

Activity: Comparing categorisation

- Two people from each group move onto the next group (remembering how they have categorised the information).
- They compare what they have done with the new group.
- They discuss what story they have to tell about the eruption.

Example 2: Categorising. Source: Anna Mercy ... continued

Plenary

- One group is selected to tell the story of the eruption. The other groups have to listen carefully, to see if anything is missing from the story.
- Other groups add additional information.
- Debriefing prompts: What kinds of categories did you find useful to use to make sense of the information? What differences did you find between your first group and your second group? Is there anything you do not understand about the eruption? (Try to elicit and answer questions.) This information came from the commentary on a programme made for schools: How reliable do you think this information is? The programme-makers had to select from a lot of information: Would you have liked more detail about anything?

(b) Cards for categorising

57 lives were lost	School children wrote letters to Harry to try to persuade him to come down	Hundreds of small fires were ignited
221 houses were destroyed	Scientists predicted that the volcano would erupt before the end of the century	An area around the volcano that was thought to be unsafe was called the 'red zone'
27km of railroad were destroyed	On 20 March the largest earthquake ever recorded at Mount St Helens occurred	People in the Valley did not hear anything when the volcano erupted
27 bridges were destroyed	The plates that rubbed together were called the Juan de Fuca and the Pacific plates	Harry would not come down from his home near the volcano because he did not think it would erupt
295km of road were destroyed	After the eruption the river was 0.5km wide	135km away, the city of Seattle was plunged into darkness
The blast went sideways out of the mountain	The river peaked at 6.5m above flood level	The scientists could not say exactly when it would erupt
On the day before the eruption people were allowed back into the red zone	Harry was killed in the eruption	Mount St Helens is 80km north of Portland and 160km south of Seattle
The eruption happened on 18 May 1980	Trees were ripped from their roots	Mount St Helens is in the Cascade Mountain range
Harry Truman lived next to Spirit Lake for 84 years	The lava flowed at 320kph	

DIAMOND RANKING FACTORS

This activity is applicable to situations in which explanation and interpretation are related to many contributory factors. Students are provided with a set of at least five or nine cards on which are written contributory causes. Students have to consider these and decide which is the most important, which are the next two most important, and so on, so that they can arrange the cards in a diamond pattern.

Diamond ranking can be used as revision or starter activities or to analyse data.

Diamond ranking as a revision activity

Example 3 illustrates the use of diamond ranking, used as a revision activity. The cards used in this activity provided minimal information.

Diamond ranking as a starter activity

Diamond ranking can be used as a starter activity to elicit students' existing knowledge, their values and to create a 'need to know'. In this activity students are provided with a decision-making context together with a set of cards on which information is given about factors to be taken into account. The task is to arrange the cards in a diamond-ranking pattern according to which factors are most important. In the subsequent enquiry work, students apply their ranked factors to data provided in order to make their decision. For example, students could rank factors to be taken into account in a possible expansion of airports in the UK. They could apply their rankings to the airports and sites being considered for expansion and determine where would be the best place for an expansion based on their priorities. Students could learn from such an enquiry that airport expansion is not inevitable; it is a political choice. Students could also learn that if airport expansion takes place the explanation of where that expansion takes place is complex and needs to take account of human agency and political power. Students would learn that explanations in geography are complex.

Diamond ranking used to analyse data

In this activity students are provided with data about the topic to be studied, together with a set of cards. Students, in groups, analyse the data and determine which factors are most important and which factors are least important. They arrange the cards in a diamond-ranking pattern.

Alternatively, students are provided with data about the topic and produce their own diamond-ranking cards either for others to use or to discuss in a plenary discussion.

Example 3: Diamond ranking. Source: Chris Holt.

(a) Procedure

Key question

What factors are important in locating an industry?

Resources

- 12 cards on factors influencing industrial location (Example 3b)
- Diagram of diamond ranking on board (Example 3c)

Starter

- Teacher introduces class to the activity as a revision exercise.
- Teacher challenges class to think of reasons why industries are located where they are.
- Students divided into groups of three.

Activity: Diamond-ranking factors generally

- Each group of three students is given 12 cards.
- Students discuss which three cards they do not want to use and put these on one side.
- Students discuss the factors on the remaining 9 cards and put the most important factor at the top.
- They put the least important factor at the bottom.
- They arrange the other cards in a diamond pattern.

(At this stage, students might start saying that it depends on the industry. Ask them to explain what they mean and keep their comments in mind until later. Ask them to do what they can generally.)

- Each group decides which two students should move on to the next group. The two students study their arrangement of cards and remember them.
- Two students from each group move on to the next group.
- The two students look at how the next group arranged the cards, looking for similarities and differences.

- The group discusses reasons for placing the factors where they did.

Interim plenary

Debrief prompts: Which cards did you reject and why? Which card(s) did groups put at the top? Why do you think this factor is important? Which card(s) did groups put at the bottom? Why do you think this factor is least important? What kinds of differences were there between the groups? What problems did you have in arranging the cards? Why might different factors affect different industries differently?

Activity: Diamond-ranking factors for particular industries

- Students remain in groups of three.
- They are given an example of a specific industry and asked to re-arrange their cards to apply the factors to this example.

(They can either be given several different examples in succession, or different groups can be given different examples.)

Summary plenary

Debrief prompts: Are there any factors that were important for all industries? Are there any factors that influence location, but are less important generally? Are there any industries with particular locational factors? How important are economic factors? How important are environmental factors? How important are social factors? What have you learned from doing the diamond-ranking activities? How did you make your decisions in your group? What kinds of differences were there between groups? In the world outside the classroom, would different groups of people rank these differently for the same industry? Why? How do you think differences between groups are resolved?

(b) Factor cards

Close to universities and research facilities	Enough room to expand	Geographical inertia
Pleasant surroundings nearby	Natural routes like rivers, valley floors	Government policies e.g. reduced taxation, incentives, grants
Good transport links	Close to the market	Good supply of labour
Close to raw materials	Suitable land for building	A good power supply

Example 3: Diamond ranking. Source: Chris Holt ... continued

(c) Diamond-ranking pattern

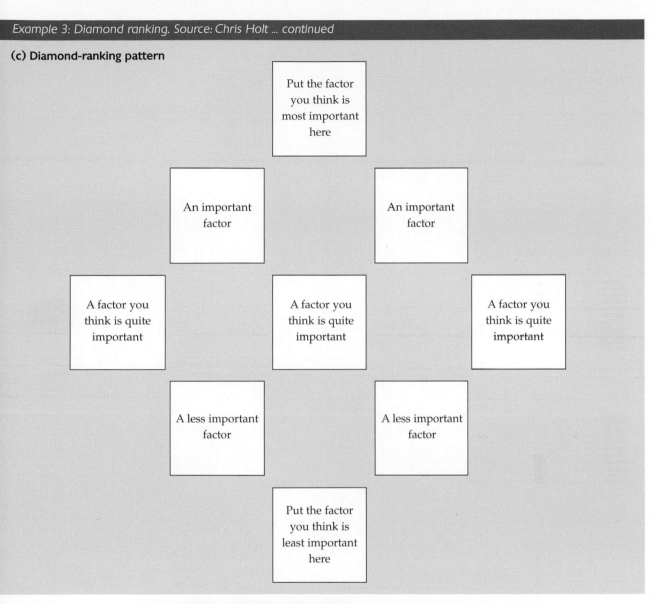

COMPARING AND CONTRASTING

If students are to learn that what happens in the world is not rule-bound and inevitable, it is useful for them to investigate similar events which have had different impacts in different places. There are many possibilities, e.g. the effects of earthquakes of similar magnitude, the impact of closure of coalmines in different areas, the effects of flooding in different parts of the world. A comparative investigation asking *why* encourages students really to get inside the question 'Why is it like this?' and to recognise the complexity of cause and effect in geography. They need to identify similarities and differences and to consider which of the differences are significant. It is a challenging activity requiring considerable thought. Example 4 shows a note-taking frame used to help students select appropriate information from a video about the Kobe earthquake. This was the first of three lessons designed to enable year 9 students to answer the question: Why did more people die in the Gujarat earthquake than in the Kobe earthquake? In the following lesson they used a similar note-taking frame to select information from the internet about the earthquake in Gujurat, India. In the third lesson they used their notes to write an account explaining why they thought more people had died in the Gujurat earthquake.

Example 4: Kobe earthquake, note-taking frame. Source: Jeanette Shipley.

THE KOBE EARTHQUAKE!

General Information

Date:

Time:

Size of the earthquake

Any aftershocks?

Where is Kobe?

What is Kobe like?

What was the cause of the earthquake?

How many people died/were injured?

Effects of the earthquake

What type of buildings collapsed? How? Why?

Why did so many people die?

Think about:
• Location of the epicentre
• Density of the population
• Quality of the building design
• Role of the emergency service
• Role of the Japanese government
• Wealth of the country

CLUES

Another strategy for enquiries focused on explanation is to provide students with initial information about a process or a theory. Students are then provided with information to look for evidence for the process or theory.

Example 5a shows the route of a walk in school grounds to look for clues about different forms of weathering. Year 7 students had to make observations at each of eight sites. They then worked in pairs to discuss and decide what the cause of the weathering was at each site. They used an observation sheet (Example 5b) to record what they saw and their explanation for it.

Another way of using the 'clues' strategy is to provide students with initial information about factors affecting climate in the UK or in Europe. Students then study current weather maps to look for evidence for each of the factors. Students annotate the maps with their observations. It is possible that they will find some evidence supporting the theory (e.g. decrease in temperature with latitude) and other evidence not supporting the theory (e.g. the east could be drier than the west on a particular day). The activity would make them look at evidence and could make them aware of the differences between weather and climate.

Example 5: Clues: Weathering walk. Source: Jeanette Shipley.

(a) The route

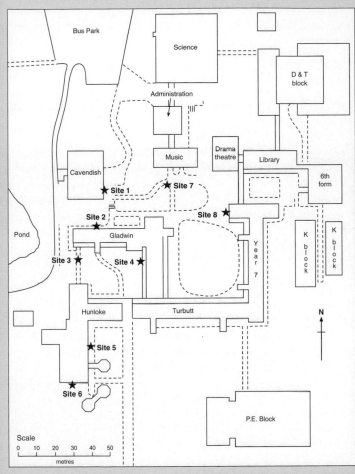

Bus Park
Science
D & T block
Administration
Drama theatre
Music
Library
6th form
Cavendish
★ Site 1
★ Site 7
Site 2 ★
Site 8 ★
Pond
Gladwin
Year 7
K block
K block
Site 3 ★
Site 4 ★
Hunloke
Turbutt
N
★ Site 5
★ Site 6
P.E. Block

Scale
0 10 20 30 40 50
metres

(b) Worksheet

Weathering walk

Name...

- In each box **describe** what you see at every site.
- Use the school map to locate each site.
- Fill in the 'type of weathering' in the classroom.

Site 1:
Wooden slat wall C1 exit of Cavendish

Cause:

Site 2:
Bottom part of wall on the north side of G2

Cause:

Site 3:
The steps by the girls toilets – Gladwin

Cause:

Site 4:
Outside G6 – pavement slabs around the tree

Cause:

Site 5:
The wall by H11

Cause:

Site 6:
South end of Hunloke – path by the fire escape

Cause:

Site 7:
The path under the tree on the south side of the music block

Cause:

Site 8:
Brick end wall, year 7 by the yew tree

Cause:

Figure 2: Literacy support for enquiry work with a focus on explaining.

Students who need additional support for written work could be provided with the following:

1. Geographical vocabulary and phrases related to processes, causes, effects, implications

2. General vocabulary related to change over time and to reasons, e.g. then, next, gradually, meanwhile, after this, at a later stage, because, therefore, although.

3. A few phrases and sentence starters can be devised for a particular enquiry, for example:

- There are several reasons for this ...
- One reason is ...
- The evidence for this is ...
- Another reason is ...
- One of the causes of ...
- I think the most important reason is ...
- The evidence suggests that ...
- There were several effects of ...
- One of the effects was ...
- The short-term impact of this is ...

- A longer term impact is ...
- The further implications of this are ...
- The factors affecting this are ...
- The most important factor is ...
- One explanation for this is...
- Another explanation for this is ...
- This meant that ...
- The people most affected by this were ...

LITERACY SUPPORT

Literacy skills related to explaining can be developed by creating a classroom environment in which explanation and interpretation is valued. This can include:

- encouraging students to give reasons for what they say
- providing students with opportunities to listen to, to see and to read written explanations
- incorporating in enquiry work a need to explain the findings orally or in writing
- providing opportunities for all kinds of written work in which explanations are important, e.g. accounts, leaflets explaining landscape or townscape features, newspaper reports
- encouraging the use of appropriate geographical terminology for processes
- encouraging students to compile their own glossaries or personal geographical dictionaries
- providing, for students who need it, additional support in the form of lists of vocabulary, connective words and sentence starters (Figure 2).

'When the content of geography lessons consistently fails to incorporate a values dimension, it may be accused of moral carelessness, because it fails to prepare students mentally and emotionally for an uncertain and unpredictable future in which their (and their teacher's) grasp of change may at best be described as "fragile". Values education sessions therefore provide a formal vehicle that enables moral development to take place'

(Lambert and Balderstone, 2000, p. 290).

INTRODUCTION

Almost all aspects of geography are affected by human decision making. In studying geography we need to take account of the attitudes and values that informed such decisions and of the power relations involved. The Schools Council 16-19 route for geographical enquiry included the investigation of values as an integral part of enquiry. There can indeed be a place for values enquiry in relation to almost all topics studied in geography at key stage 3. There is also a case, however, for developing enquiry work in which the focus is concentrated on attitudes and values. It prevents the all-important consideration of attitudes and values being dealt with cursorily.

This chapter firstly explores what is meant by attitudes and values. It then outlines the kinds of questions addressed by values enquiries and the opportunities they provide for learning. It suggests ideas for classroom activities which focus on attitudes and values.

ATTITUDES AND VALUES

There is a distinction between 'attitudes' and 'values'. Basically, attitudes indicate whether people are for or against something while values are about what is of enduring importance to people or organisations. It is worth expanding on these quick and easy definitions.

Attitudes are feelings, viewpoints, or opinions held by individuals, groups or organisations about an issue. Issues can range from small-scale local issues such as the redevelopment of a local park to global issues such as the preservation of biodiversity. Attitudes can include:

- Feelings and opinions *for* something
- Feelings and opinions *against* something
- An *indeterminate* viewpoint, either undecided or neither for nor against.

Attitudes about something can shift and change as people come into contact with other views. Values, on the other hand, are more deep-rooted, persistent beliefs about what is really important to people and organisations. The values that underpin attitudes about most geographical issues can be categorised as:

- social values (about the good of people and communities)
- economic values (about money and wealth creation)
- environmental values (about conserving the environment and about sustainability)
- moral values (about what is ethically right or wrong, just or unjust).

For example, an industrial establishment might have an attitude in favour of something, but the most important underpinning value is likely to be related to economic values: industries, unless they can rely on subsidies, have to make money. On the other hand, an environmental organisation might be against a new development, because it threatens a rare ecosystem, an attitude underpinned by environmental values.

It is rarely as simple as this. Often there will be conflicts in values, not only between people, but also between different types of values. For example, someone might value both full employment and the conservation of a threatened environment. For this person, the proposal for a new airport in an area of high unemployment might lead to a clash between environmental and social values and this makes his/her attitude towards it problematic. An organisation might be concerned about both profits and about the environment. The fact that a person or organisation has strong values does not mean that attitudes to important issues are necessarily straightforward.

Figure 1: A framework for learning through enquiry: values and attitudes.

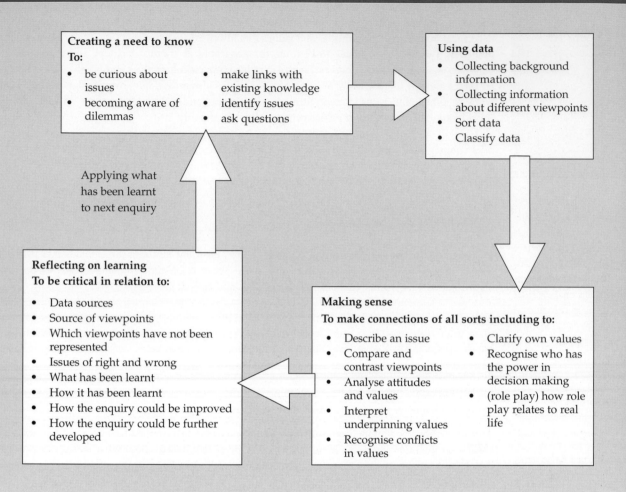

Enquiries that focus on values are essentially about conflicts of values. It is, of course, important to investigate attitudes towards an issue and the evidence in favour of different viewpoints, but issues are not resolved by rational argument about the facts of a case. Issues are debated not only in relation to the facts but also through arguments about what is believed to be most important (Figure 1).

KEY QUESTIONS

Enquiries that focus on values need to include questions addressing both attitudes and values and might include some of the following:

- What is the issue? What are the basic facts?
- What attitudes do different people/organisations have about this issue?
- Why do they think like this? What evidence is there to support their views?
- What is important to them? What are their underpinning values?
- What conflicts in values are there?
- Who has the power to decide?
- What do I think about the issue? (What is my attitude?)
- What is important to me? (Which values are important for me?)

- What do I think should be done?
- What could I do about it?

LEARNING OPPORTUNITIES

Enquiries with a focus on values provide opportunities for students to:

- know that people have different attitudes to issues
- become aware of a range of viewpoints
- develop skills of analysing evidence
- become aware of underpinning values
- become aware of conflicting values
- become aware that decisions on issues can be influenced by perceptions
- become aware of power relations in decision making
- clarify their own attitudes and values
- express their own views
- develop understanding of complex issues.

Depending on how the enquiry is developed, there may also be opportunities for students to:

- develop speaking and listening skills
- ask questions of others
- answer questions using evidence
- develop numeracy skills
- develop literacy skills, including increasing familiarity with discussion and persuasion text types.

WHERE DO VALUES ENQUIRIES FIT IN?

There is scope for developing enquiry work with a focus on values on every theme and country studied in the key stage 3 geography national curriculum, for example:

- Countries: What should the priorities of this country be for its future development (or an aspect of development)?
- Volcanoes or earthquakes: How should limited funds be spent in making preparations for an earthquake or volcanic eruption?
- Coastlines/rivers: How should a stretch of coastline/river be managed?
- Migration: Where and how should asylum seekers be housed?
- Settlement: Should a new airport be developed in this area?
- Supermarket (or other economic activity): Should the supermarket be built in this location?
- Rainforest: What should this area of forest be used for?
- Energy supplies: What types of energy should we plan to be using in the UK in 20 years time?
- Development issues: Which development projects should we fund?
- A major dam project, e.g. the Three Gorges Dam in China: It needs more money: should the World Bank support it?
- Tourism: Where shall this group (e.g. particular family, the whole class) go on a holiday?

ACTIVITIES

The following classroom activities explore the attitudes and the underpinning values of those with different views of a geographical issue:

- stakeholders
- hot-seating
- role play

STAKEHOLDERS

In this activity the students' task is to explore the question: Who would have an interest in this issue and why? Listed below are suggested stages for a stakeholder activity.

Stage 1: Students are introduced to the issue, e.g. should permission be given for the leisure centre to be built?

Stage 2: Paired or group work: students study data (e.g. photographs, video, newspaper extracts) provided on the issue and decide which people and organisations might be for the development and who might be against the development. All those for or against are the 'stakeholders' in the issue and they have an interest in whether it goes ahead.

Stage 3: Paired or group work: students produce a spider diagram showing the 'stakeholders'. They then devise a way of showing whether each stakeholder is for or against the development and make a key.

Stage 4: Interim debriefing: students share their ideas of who might have an interest in the issue. Students add to their diagram if they have omitted key stakeholders.

Stage 5: Paired or group work: students discuss why each stakeholder would be for or against and make a list of different types of reasons.

Stage 6: Debriefing: Students feed back to the whole class the different kinds of reasons why people might be for or against the development. These are discussed and the class try to agree a list of underpinning values. Who should decide whether it should go ahead? How are such decisions made? What influences such decisions?

Stage 7: Follow up: written work either in tabular form (listing stakeholders, for or against, and why) or extended writing or in writing expressing a personal reasoned view on the development.

HOT-SEATING

Hot-seating is an activity in which students are put on the spot in the 'hot seats' to speak to the rest of the class or to answer questions in role. The best classroom arrangement for hot-seating is a circle of chairs in which everyone can see each other easily. Listed below are suggested stages for a hot-seating activity.

Stage 1: Students are introduced to the issue and to the key questions being explored in such a way as to 'create a need to know'.

Stage 2: Group work: students investigate the issue from the viewpoint of a particular group of people. They collect information from data provided (either data common to the whole class or particular to their role or both).

Stage 3: Hot-seating plenary: the class sits in a circle, if practicable, and groups take it in turn to sit in the hot seats. Ground rules for speaking and asking questions need to be established.

Stage 4: Each group introduces itself briefly and then answers questions from the other groups (asking questions in role).

Stage 5: Debriefing: What are the main points about the issue? Why do people have different attitudes towards it? Which questions were difficult to answer in the hot seats? Why were they difficult? Which viewpoint came across best in the hot seats? What do you think about the issue now?

ROLE PLAY

The public or private meeting is an excellent scenario for classroom values enquiries. Role-play lessons are usually exciting, stimulating for students and highly memorable. More than in any other types of enquiry work, students are likely to go out of role-play enquiries still discussing the issues. The involvement of students in speaking and listening encourages them to continue thinking about the issues. Whereas typical lessons are dominated by teacher talk, role-play lessons are dominated by student talk and by high levels of student engagement.

Figure 2 suggests the types of decisions which are suitable for public meeting role plays. A checklist of what needs to be considered when preparing public meeting role plays is set out in Figure 3.

Figure 2: Issues suitable for public meeting role play.

Choosing one of several sites

In this category, there would be roles for people advocating each particular site.

Examples of suitable issues include:

- Which venue should be chosen for a major sporting event, e.g. Olympic Games 2012?
- Which of these towns should be given city status?
- Which is the best place to locate a particular new factory?

Choosing one of several policy options

In this category there would be roles for those with some expert knowledge about the different options and also roles for those with an interest in the different options. Possibilities include:

- Which of the proposals for this coastline should we accept?
- Which of the proposals for the UK's future energy use should be adopted?

Making a yes or no planning decision

In this category there would be roles for different groups with an interest in the issue. Possibilities include:

- Should a supermarket be built on this site?
- Should domestic waste be incinerated to produce electricity?
- Should this bypass be built?

Allocating scarce resources

In this category there would be roles for people advocating different proposals, each of which needs funding. The decision could be either about how much to allocate to each proposal or which proposals to support and which proposals not to support. Possibilities include:

- Which of these possible development projects (e.g. for Oxfam) do we support?
- How should this town spend limited money to prepare for the next earthquake?

Figure 3: Checklist for planning a public meeting role play.

Select a suitable issue

- Decide the key questions and the type of decision to be made.

Decide on roles

- Decide what roles are needed, including:
  - Chairperson
  - Groups representing different viewpoints
  - People to make the decision
  - Other roles if needed

Prepare resources

(Data could be provided in the form of text, maps, statistics, brochures, advertisements.)

- Prepare background information for all students
- Prepare information for particular groups and possibly for individuals within groups
- Prepare name cards

Decide on procedure for role play

- Produce agenda for the public meeting for handing out or for display
- Produce, if required, a note-taking frame to be completed during the meeting

Prepare the classroom

- Re-arrange tables or desks
- Put up any necessary maps, display information and agenda
- Set out name cards

Plan the debrief session

- Prepare some key questions for debriefing

Plan follow-up work

- Decide whether written work is to be required. If so, write instructions.

Fulin, the lower part of which will be flooded by the Yangtze River.

Role-play can offer students the opportunity to become involved in the whole process of enquiry in all its essential aspects (Figure 3). Often there are mini-enquiries during a public meeting when students think of questions to ask on the spot and in turn they have to answer questions using the data provided. Role plays can also make students aware of how some voices are heard more than others and about how decisions about issues can relate to issues of power and vested interests as much as to attitudes and values.

Example 1 sets out a procedure and some resources for a role play related to the construction of the Three Gorges Dam on the Yangtze River in China. Although the project is going ahead and the dam has been built, there is still need for additional funding. The need for additional funds provides the scenario. As presented in this book, because of the vested interests of those making the decision, it is likely that the project would be supported. This role play does not replicate a likely meeting, but it does encourage a discussion of the environmental, social, economic and human rights issues related to the building of the dam.

Example 1: Three Gorges Dam Project – role play.

(a) Procedure

Key questions

- What are the key features of the Three Gorges Dam Project?
- Why was the Three Gorges Dam built?
- What are the advantages of the project? Who gains?
- What are the disadvantages of the project? Who loses?

Resources

- Atlas map of China
- Map of project area (can be found at www.probeinternational.org)
- Background information on project (Example 1b)
- Role cards (Example 1c)
- Note-taking frame (Example 1d)
- Photographs (can be found on Google.co.uk images)

List of roles

- Chairperson
- International team who will make the decision
- Chinese government (*for*)
- Local people who think they will benefit from the project (*for*)
- Local traders and industrialists who think they will profit from the project (*for*)
- Environmental and conservationist action groups (*against*)
- Local people who think they will be worse off (*against*)
- Economists against the project (*against*)

Starter

- Introduce the scenario. The Three Gorges Dam Project needs further international investment to complete the project. Three countries are considering whether to support the demand for further investment. There will be a meeting to discuss this in which the arguments for and against the project will be voiced.
- Introduce the project (using or going through the information in Example 1b). Emphasise:
  - the scale of the project: the biggest dam project in the world, the size of the reservoir (compare with a distance in the UK)
  - the reasons for the project: to protect against flooding (give information about history of flooding); to generate electricity; to improve navigation
- Allocate roles
- Introduce note-taking frame

Activity: Preparation for role play of public meeting

Students, working in groups:

- read information on the role card
- possibly search for additional information
- prepare speeches for role play and decide who is going to say what
- prepare labels for their group to put in front of them.

Activity: Role-play of public meeting

- The chairperson introduces everyone
- Those in favour of the Three Gorges Dam project make their case (each presentation followed by one question from the international team)
- Those against the Three Gorges Dam project make their case (each presentation followed by one question from the international team)
- Five-minute interlude: each group devises questions for one of the opposing groups (as listed on role card) based

on the arguments presented

- Question time: each group asks questions and the appropriate group responds
- The international team leaves the room to consider their decision
- Interim debrief while the decision is being made: Which of the arguments for the project seemed the strongest? Which of the arguments against the project seemed the strongest? What do you think the international team will decide? Why?
- The international team presents its decision to the group.

Activity: Debriefing

- To the international team: What were the most important issues for you?
- To all groups: Do you think the decision was fair? What were the weakest of your arguments? What was the strongest argument against you? In what ways do you think this role play is realistic? What made the role play difficult for you? What other information would you like to have had? What have you learnt about the Three Gorges Dam project? Who gains most from the project? Who loses most from the project?
- Individual votes: If you were not in role, would you vote in favour of investing in the project or not? Count the votes. Note whether the votes are still in role.

(b) Background information

Key facts

(c) Role cards

- The Three Gorges project is named after three scenic gorges (Qutang Gorge, Wuxia Gorge and Xiling Gorge) on the Yangtze River in China
- The building of the dam at Sandouping, near Yichang, was approved in 1992 and building started in 1995. The dam will be completed in 2009
- The dam is 185m high and 2.3km long
- There are five ship locks to let ships pass from the lower part of the Yangtze River to the upper part
- The reservoir behind the dam will be 600km (375 miles) long
- The maximum height of the water above the present level of the river will be 175m.

The main reasons for building the dam

- To prevent and control the floodwaters of the Yangtze River (in the twentieth century more than 300,000 people died as a result of flooding from the Yangtze River)
- To generate 18.2 million kilowatts of electricity per year
- To improve transport along the Yangtze to Chongqing
- To supply water from the reservoir.

The impact of creating the reservoir

- The area to be flooded by the new reservoir includes: the Three Gorges, 13 cities, hundreds of villages, a large amount of farmland, 1300 factories and many ancient monuments
- A total of over 1,130,000 people will have been moved by the time the project is completed.

Local people, including farmers (for the project)	Environmentalists and conservationists (against the project)
- Some of you live in the area to be flooded by the reservoir. You have been moved to higher land in your own town. Your old flats were cramped and damp, without water supply or electricity. The new areas where you live have wide roads and modern apartment blocks with water supply and electricity. Your housing conditions are much better - Some of you live in the land next to the Yangtze River below the dam. You have always been at risk of having your home and your farmland flooded. Some of your relations drowned in recent floods and others lost everything they had. When the waters of the Yangtze are controlled you will not need to worry about flooding any more - You have received compensation for moving and feel better off than you have ever been. *When the local people against the project have made their presentation, think of a question to ask them. Be prepared to answer a question from them.*	- You are worried about pollution of the reservoir. The reservoir will flood thousands of rubbish dumps and will drown 1300 factories. The poisonous waste will include arsenic, mercury, lead, and cyanide. This will go into the reservoir. It will get into farming land and the drinking water - You are worried about all the silt in the Yangtze River. The water will move more slowly and the silt will build up. In another dam on the Yangtze River the amount of electricity produced has been reduced because of silt - You are concerned about the flooding of 1200 ancient monuments. They are an important part of Chinese history. *When the Chinese government has made its presentation, think of a question to ask them. Be prepared to answer a question from them.*

Example 1: Three Gorges Dam Project – role play ... continued

Chinese government (for the project)

You are in favour of the project because:

- It is a prestige project. It is the biggest project of its kind in the whole world

- It will bring the waters of the Yangtze River under control. It will reduce the risk of flooding from the Yangtze River, especially below Yichang. There have been disastrous floods from the Yangtze River since ancient times. During the twentieth century over 300,000 people were killed and many lost their farmlands

- It will generate a vast amount of electricity for central and eastern China. Demand for electricity is growing quickly in China. The electricity from hydro-electric power is clean and renewable. It will be possible to close down many of the coal-fired power stations. This will reduce the emission of carbon dioxide into the atmosphere. This will reduce acid rain and global warming

- The reservoir can be used to supply water to large areas of China.

Additional information

Research has been carried out into the risk of earthquakes in the area. The dam is being made of very high quality concrete. Engineers are sure that it can withstand any likely earthquakes.

When the environmentalists have made their presentation, think of a question to ask them. Be prepared to answer a question from them.

Local people, including farmers (against the project)

- You used to live in the towns and villages near the Yangtze River. Now your homes and your farmland have been flooded. You used to have land to grow your own vegetables and to grow orange and lemon trees. You have had to move to a town hundreds of kilometres away where you do not know anyone. The people there speak differently from you and they treat you badly. Some of you have been given land, but it is not enough. Some of you have to find jobs in the town. You cannot grow your own food and the food in your new town is very expensive. You have not received the compensation you were promised

- You know some people who have protested about not receiving the money. Some of them have been put in prison.

When the local people for the project have made their presentation, think of a question to ask them. Be prepared to answer a question from them.

Traders and industrialists (for the project)

- You are very excited about the project. Travel by boat to Chongqing will be much faster and safer. There will no longer be dangerous rapids in the Three Gorges part of the river. The waterway will be wider and the water level will be higher in the dry season. This means that transport will be much better. New roads and bridges are being built. All this is good for trade

- You think that tourism will increase. The scenery of the Three Gorges is dramatic. The mountains rise over 1000m on either side of the river. The scenery will still be dramatic after the level of the water rises 175m. The most important ancient monuments are being moved. Some of the higher monuments will be easier to reach. You think that the project will become one of the wonders of the world and will attract more and more tourists. More hotels will be needed. More tourist boats will be needed. It will be good for trade

- There will be a lot more jobs for everyone in the area.

When the economists have made their presentation, think of a question to ask them. Be prepared to answer a question from them.

Economists (against the project)

- You think it is risky to spend money on the project. It is expensive to transport electricity over huge distances. The electricity produced by the project will be very expensive. Not many people in China can afford this. There will not be enough demand for the electricity

- If there is the demand for electricity, there is a risk that it will not be produced. Another dam on the Yangtze has silted up so much that it only produces half of the electricity planned. There is always an increased risk of earthquakes when there are large reservoirs, because of the weight of the water. If earthquakes damaged the dam it would be expensive to repair. If the dam burst it would be a huge disaster

- You think the dam could be a target for terrorists. It will be expensive to protect.

When the local traders and industrialists have made their presentation, think of a question to ask them. Be prepared to answer a question from them.

Example 1: Three Gorges Dam Project – role play ... continued

International team (you have to make the decision)

You represent three countries.

Canada

In the past your country has supported the project. Your industries supply the project with:

- Cement plants
- Turbine generators
- High voltage electric equipment.

The Export Development Corporation has given loans for the project to the People's Construction Bank of China.

You are likely to support the project because Canadian companies have invested a lot of money in it.

Take this into account when you listen to the arguments.

You have to decide whether your country supports further investment in the project.

Japan

Your country has some investment in the project. Your industries have supplied the project with:

- Some construction equipment
- Steel plate.

Your country suffers from acid rain because of China's coal-fired power stations. This might reduce when this is replaced by hydro-electric power.

You do not have a lot of investment but you might be concerned about acid rain.

Take this into account when you listen to the arguments.

You have to decide whether your country supports further investment in the project.

United Kingdom

You do not have much investment in the project. Your industries produce:

- Control systems for dredging vessels.

You may want to increase your trade with China.

Some people in your country are worried about human rights in China.

Take this into account when you listen to the arguments.

You have to decide whether your country supports further investment in the project.

(d) Note taking frame

Main focus	People for the project	People against the project
Environmental issues	Chinese government	Environmentalists and conservationists
Social issues and human rights	People who live locally, including farmers	People who live locally, including farmers
Economic issues	Traders and industrialists	Economists

Example 2a outlines the procedure used in a year 8 class to investigate a topical controversial issue in Sheffield: the expansion of Bernard Road incinerator. Students were able to study local newspaper cuttings, but they obtained most of their information from the internet. They then used guidance sheets written specifically for each of the roles. The role play was lively, with all students fully involved. The students wrote about their own views for homework (Example 2b). The teacher was impressed with the amount and the quality of the writing compared with what these students usually wrote. The writing was more engaged and involved in the subject matter and benefited from the discussion and thinking taking place in the lesson.

Example 2: Bernard Road incinerator. Source: Jenny Allen.

(a) Procedure

Key questions

- Should permission be given for a larger incinerator to be built in Sheffield?
- What are the arguments in favour of Bernard Road incinerator?
- What are the arguments against Bernard Road incinerator?

Resources

- Photograph of Bernard Road incinerator (can be obtained from the internet)
- Guidance sheets for each role (providing information on how to access internet information through Google.co.uk)
- Access to the internet.

List of roles

- Chairperson (teacher)
- Decision maker (another teacher)
- Residents Against Bernard Road Incinerator (RABID) (*against*)
- Green Party (*against*)

- Greenpeace (*against*)
- Sheffield Heat and Power Company (*for*)
- Sheffield Council/Onyx (*for*)
- Other Sheffield residents (*for*)

Lesson 1: Preparation for the public meeting

- Students are introduced to the issue
- Students are allocated roles
- Students research the issue on the Internet using guidance sheets.

Lesson 1: The public meeting

- Presentation of arguments for and against the proposal to build a larger incinerator
- Students question each other
- Decision is made.

Homework

- Students write a report about the issue (Example 2b)

Note: Full details have not been included because of constant changes to the websites used.

(b) Students' written work following the role play

Note: Original spelling and punctuation have been retained, but all names have been changed.

Calvin

I am against the proposal for the new incinerator because: The new incinerator will cause carbon dioxide and toxic fumes. It will release thousands of chemicals into the air. People are far more likely to get cancer if the live near it. It pollutes rives and damages the ozone layer, it causes global warming and causes acid rain. The ash from the fire will still have to be taken to landfill sites. Heavy lorries will have to go to the incinerator which will cause extra traffic. People don't want to see a big chimney form their house. It causes noise pollution when it starts up. The fumes cause dioxins. If the rubbish is being burnt it isn't being recycled. The new incinerator is in an inner city area which could be used for community purposes. The £28 million which will be spent on the incinerator could be spent on other things like public transport or roads. The government should invest in sustainable types of energy such as wind power or hydro-electric power.

Matthew

I am against the incinerator because of the location of it been in the middle of the city, and we don't know what effects it could have on our health in future.

It would also be a problem because we don't know what effects it could have on wild life, and it could cause smog.

It would be a problem for people who live in the area because of the noise it makes, and if it is bigger it means it will make more noise and more stress.

Lisa

I disagree with Bernard Road incinerator because of all the pollution which comes out of it. The smoke which comes out of it from some of the rubbish burnt is very toxic and caused damage to the air which has an effect on the ozone layer and to us. It damages our lungs which causes dieasus.

There has been incinerators from quite a long time it is an easy way to dispose of rubbish and safer than using a landfill because the land is very toxic for a long time after it has been filled with gases from rubbish. More recycling would cut down the amout of rubbish ... Maybe a safer incinerator could be build with filters to stop air pollution and not near where people live.

| Figure 4: Literacy support for enquiry work with a focus on values. | LITERACY SUPPORT |

Students who need additional support for written work could be provided with the following:

1. Geographical vocabulary and phrases related to the issue being studied

2. General vocabulary related to attitudes and values, e.g. opinion, point of view, viewpoint, standpoint, argument, belief, on the one hand, on the other hand, however, alternatively, for example

3. A few phrases and sentence starters can be devised for a particular enquiry, for example:

- Some people would argue that …
- _____ thinks that …
- One of the reasons for thinking this is …
- Other people would argue that …
- _____ thinks that …
- The arguments for this point of view are …
- The implications of this are …
- If this were to happen then …
- The decision about this issue will be made by …
- The people with the most influence in this issue are …
- The people with the least influence are …
- On balance I think …
- The people who will be affected by this issue …

LITERACY SUPPORT

Literacy skills related to values can be developed by creating a classroom environment in which attention is given to attitudes and values and the reasons for them. This can include:

- investigating issues for which there are different viewpoints and in which there is a need to understand the underpinning values
- providing students with opportunities to listen to, to see and to read the opinions of others
- incorporating in enquiry work a need for students to express their own opinions orally or in writing
- encouraging students to give reasons for what they say
- providing opportunities for all kinds of written work in which values and attitudes are important, e.g. discursive accounts of issues, newspaper reports of issues, letters expressing an opinion, leaflets or posters promoting a cause
- providing, for students who need it, additional support in the form of lists of vocabulary, connective words and sentence starters (Figure 4).

'We should recognise the
impact that social surveys
and other related statistics
have on understandings of the
social and cultural world'

(Shurmer-Smith,
2002, p. 109).

INTRODUCTION

A considerable amount of the secondary data encountered by students in geography textbooks and in the media originates from questionnaire surveys. If students are to become more fully aware of how this kind of knowledge is constructed, then it is advantageous for them to become involved in the process of using questionnaire surveys to answer geographical questions.

Although questionnaire surveys are usually used in school geography in relation to fieldwork, they can also be used for enquiry work taking place mostly within the classroom. This chapter sets out the geographical questions that can be investigated using questionnaires, and their relevance to the national curriculum. It discusses the opportunities for learning offered by survey work as well as the limitations of such work, and concludes with some classroom examples.

KEY QUESTIONS

Questionnaire surveys can be used to answer a wide range of geographical questions related to both facts and opinions, for example:

- What? Where? When? How often? Why?
- What do you think? Do you agree or disagree?
- What is most important to you?

Questionnaires can provide information about:

- the person answering the questionnaire, e.g. age
- behaviour, e.g. shopping habits
- opinions, e.g. attitudes towards an issue
- existing knowledge, e.g. Which EU countries can you name?

Eight common question types are shown in Figure 1. Where students construct questionnaires themselves, then it is useful for them to be introduced to two or three of these question types. However, if teachers construct the questionnaires for the enquiry, then again it is useful to select a limited range of question types so that students can become familiar with them. Generally, closed questions and 'fill in the blank' questions are the most suitable for key stage 3 questionnaires because they are much easier to analyse. Closed questions invite respondents to tick, cross or circle something already on the questionnaire (question types 1-5). 'Fill in the blank' questions (question type 6) asking for only one number are also easy to analyse. Open questions invite respondents to answer a question in a word or phrase, a list or comment (question types 7 and 8). Open questions have the advantage of allowing for unanticipated responses, but they are much more difficult to analyse.

WHERE DO QUESTIONNAIRE SURVEYS FIT IN?

Many different aspects of key stage 3 geography could be studied in part through questionnaire surveys. The examples given after each topic listed below indicate the range of questions that could be asked. Each question would need to be developed to be suitable for use in a questionnaire.

- Perceptions of places (e.g. What do you associate with the following country? Which countries would you most like to visit? Why? Which countries would you least like to visit? Why?)
- Weather (e.g. weather diary)
- Holidays (e.g. day trips/longer trips: Where? Why? For how long? Type of accommodation? What holiday activities preferred?)

Figure 1: Eight common types of questions for use in questionnaires.

1. Choosing *one* answer from a list of options

e.g. Have you been abroad in the last 12 months? Yes/No

e.g. Do you live in: (categorical data)

    A terraced house   A semi-detached house

    A detached house   A flat   A caravan   Other

e.g. How old are you? (grouped data)

    Under 16   16-40   41-65   over 65

2. Choosing one or more answers (or none) from a list of options

e.g. Which of these EU countries have you visited?

Austria	Belgium	Denmark	Finland	France
Germany	Greece	Ireland	Italy	Luxembourg
UK	Portugal	Spain	Sweden	The Netherlands

3. Ranking items

e.g. Which of the following are most important to you in choosing a holiday?

Indicate your order of preference with 1, 2, 3, etc.

Holiday feature	Order of priority
Weather	
Opportunities for sport and outdoor activities	
Night life	
Interesting towns, museums, etc., to visit	
Food	
Attractive scenery	

4. Rating using the Likert scale

The Likert scale, developed by Rensis Likert in 1932, invites views on a list of statements, for example:

Statement	Strongly agree	Agree	Neither agree nor disagree	Disagree	Strongly disagree
Geography should be compulsory until the end of year 11					

5. Semantic differential

e.g. Smith Street is:

Dirty	1	2	3	4	5	Clean
Interesting	1	2	3	4	5	Boring
Noisy	1	2	3	4	5	Quiet
Unsafe	1	2	3	4	5	Safe

6. Fill in the blank

e.g. How many times have you been to the cinema in the last month? _____

7. Open list

e.g. Which countries have you visited?

8. Open response

e.g. What changes would you like to see in your town in the next five years?

- Visiting relations (e.g. Which relations? Where do they live? How long are the visits? What occasions? What do you like/dislike about visiting?)
- Journey to school (e.g. Distance to be travelled? Mode of transport? Time taken for journey? Likes and dislikes about journey?)
- Shopping (e.g. Where do you shop? Which shops? How often? What for? What features of shopping centres do you like/dislike?)
- Use of water in the home (e.g. What is water used for? How much is used?)
- Moving home: migration as a common experience (e.g. list the different places in which you have lived. Why have you moved house? What do you think of the area you last lived in? What do you like/dislike about your new area?)

LEARNING OPPORTUNITIES

There are many advantages of using questionnaire surveys as an enquiry strategy at key stage 3. Questionnaire surveys can be used to:

- involve students in the whole process of geographical enquiry from devising questions through to reaching conclusions and evaluating the work
- develop a wide range of data-handling skills
- introduce students to techniques of questionnaire design and graphical representation
- provide a purposeful context for developing ICT skills
- enhance geographical understanding
- make students aware of how knowledge is constructed through surveys
- make students aware of the limitations of findings from questionnaire surveys.

LIMITATIONS OF QUESTIONNAIRE SURVEYS

In spite of their many advantages, questionnaire surveys have some limitations in the context of classroom-based work at key stage 3.

- There will be a limit on how frequently friends, neighbours, relations, teachers or other possible respondents are willing to take part in surveys
- The sample is likely to be too small to generate valid generalisations; the findings could lead to misunderstandings
- The sample is unlikely to be random so the findings could be biased
- The process from generating questions to reaching conclusions can be time-consuming
- The extent to which human behaviour can be studied through the collection of quantitative data and attempts to generalise is a debatable issue. There is an increasing emphasis in geography in higher education on the use of qualitative approaches to the study of human geography. However, our understandings of the world are influenced by the widespread use of survey data, both specifically in geographical education and also generally in the world outside the classroom. If we are to encourage students to begin to deal with this kind of information critically, then they need to begin to understand the ways in which data is collected and for what purposes. They need to develop understanding of the selective nature of what is often presented as objective information.

WAYS OF USING SURVEYS

In classroom enquiry at key stage 3, questionnaire survey data can be:

- collected within the classroom from the students
- collected during homework by students
- collected during fieldwork
- provided by the teacher.

Examples of these uses are provided below.

USING QUESTIONNAIRES FOR DIAGNOSTIC ASSESSMENT

Questionnaires can be used as an initial activity with the purpose of finding out what students already know about a topic. This diagnostic use of questionnaires can reveal students' existing knowledge, their misunderstandings and their attitudes. It can also enable students to become familiar with different types of survey questions.

In this use of questionnaires, teachers design the questionnaires and students complete them in class. Responses are collated either during class discussion or after the lesson for later use or to inform planning of later lessons. Example 1 shows a questionnaire survey used to find out students' existing understandings of environmental problems before the start of a unit of study.

Example 1: Survey used for diagnostic assessment. Source: Rachel Gregory.

1) What do you think is meant by the term 'environmental problem'?

_____

_____

2) List three **global** environmental problems:

a) _____

b) _____

c) _____

3) List three **local** environmental problems:

a) _____

b) _____

c) _____

Acid rain

1) What is acid rain? _____

2) What causes acid rain? _____

3) What are the effects of acid rain? _____

4) Who do you think is affected by acid rain? Does it affect you? _____

_____

5) Where is acid rain a problem? _____

6) How do you know about acid rain? _____

_____

Note: The questionnaire included similar sets of questions on global warming and water pollution.

USING QUESTIONNAIRES DESIGNED BY STUDENTS

The advantage of getting students to design their own questionnaires, as a class or individually, is that this enables them to be involved in the whole enquiry process (Figure 2). Students learn best about identifying suitable questions and designing a questionnaire by doing it for themselves, and they can learn from their mistakes. They can develop a wide range of skills and techniques and learn to make their own choices. Through active involvement in the whole process, students can become aware of the selective nature of knowledge and of the possibilities of findings being misleading for various reasons. They can become aware that 'factual' information collected through surveys is a selection of reality, highly dependent on the nature of the questions asked, the nature of the sample, and the interpretation of the responses. Example 2 describes the use of a questionnaire survey in an enquiry that formed part of a year 8 assessment activity. Students were motivated by this work and learnt a lot about the process of investigating. They were also made aware of some of the potential problems of surveys, particularly in relation to the sample of respondents and the validity of their responses.

Figure 2: The eight stages of using and evaluating questionnaire surveys.

Stage 1: Establishing the scope of the survey

The teacher introduces the theme or issue which is to be the focus of the enquiry. Students identify aspects of the topic that could be investigated and these are pooled and discussed. The class discusses who should answer the questionnaires, e.g. another class in the school or people living in the same household as the students, or targeted at a particular age group.

Stage 2: Devising questions

The teacher introduces students to three or four types of questions suitable for questionnaires (Figure 1). If this it the first time students have devised their own questionnaires, it would be useful to discuss examples of each type of question in relation to a familiar topic, e.g. school subjects, leisure activities.

Students work in pairs or groups devising questions on the theme to be investigated. These are shared as a class, discussed and suitable questions selected for the final questionnaire. The sequence of questions is discussed.

Stage 3: Producing the questionnaire

The teacher produces copies of the questionnaire.

Stage 4: Collecting responses

Students find appropriate people to complete their questionnaires.

Stage 5: Collating the data

The data from the surveys needs to be combined. This can be done by students entering their own data on a computer spreadsheet or database, or by teachers.

Stage 6: Presenting the data

Students study the collated data. They discuss possible ways of presenting the data graphically. They choose appropriate ways of presenting the data, possibly using ICT.

Stage 7: Interpreting the data, reaching conclusions

Students describe what the graphs show, identifying significant features, and they make generalisations. Students summarise the main findings of the survey. The findings could be presented to a wider audience: in a display or as a report for those who helped in the survey.

Stage 8: Evaluating the survey

The findings are discussed critically. Does the data answer the questions well? Could the data be misleading in any way?

Example 2: Energy survey.

(a) Procedure

Key question

How does energy use 50 years ago compare with energy use today?

Lesson 1

- The teacher introduced the theme of energy, wondering what changes there had been in the last 50 years.
- The teacher introduced the idea of using a questionnaire survey, and of four different question types: choosing one answer from a list of options; choosing to agree or disagree with a given statement; an open list; an open response.
- Each question type was illustrated with an example and discussed.
- Students worked in groups to devise questions.
- Each group chose their best questions.
- Best questions were put on the board and discussed.
- The class discussed who could be asked to complete the questionnaire.

Before lesson 2

The teacher used the questions to produce a questionnaire.

During lesson 2

- Questionnaires were handed out, one per student.
- The class discussed how they were to be completed.

Before lesson 3

- Each student took one copy of the questionnaire home to be completed by someone who could remember 50 years ago.
- Students handed in the completed questionnaires.
- The teacher entered the data into a spreadsheet and duplicated this for the class.

Lessons 3 and 4

Unfortunately with no access to computers.

- Discussion of assessment task sheet (Example 2b)
- Support for how to get started (Example 2c)
- Discussion of use of graphs
- Students worked independently, writing the report and producing graphs

(b) Assessment task sheet for energy survey

What was the main type of energy used 50 years ago? What is the main type of energy used today?

Follow this sheet to complete your assessment fully. Tick the boxes to show you have finished each section. Good luck!

Introduction

- ❑ What are you aiming to do?
- ❑ How did you go about answering the questions above?
- ❑ What do you expect to find?

What type of energy was used 50 years ago?

- ❑ What did you expect to find out?
- ❑ What do your graphs show?
- ❑ What did you find out from the survey?
- ❑ Are the results what you expected? Did anything surprise you?

What type of energy is used today?

- ❑ What are the main types of energy used today?
- ❑ Do we use any different types of energy today from 50 years ago?

Conclusion

- ❑ Have you answered the initial question successfully?
- ❑ Is there anything else you could have done to answer the question?
- ❑ What is your main finding from this enquiry?
- ❑ Did you find out what you expected? If not, what else did you find out?

Example 2: Energy survey ... continued

(c) Energy survey: phrase and word banks

Phrase bank

For this assessment I am aiming to ...

The aim of this piece of work ...

I answered the question by ...

I expected to find ...

A questionnaire was devised by ...

With the questionnaire responses I ...

The responses to the questionnaire raised some key points ...

The graphs showed that ...

On one hand ...

On the other hand ...

I was surprised by ...

I did not expect to find ...

If I were to complete the questionnaire about energy today ...

The main types of energy used today ...

The energy that we use today is very different from energy use 50 years ago ...

If I was to complete this assessment again ...

My main finding ...

In conclusion ...

Overall ...

Word bank

Coal	Non-renewable
Oil	Sustainable
Gas	Unsustainable
Fossil fuels	Gases
Solar	Harmful
Green electricity	Carbon dioxide
Wind	Environment
Wave	Analysis
Renewable	

USING QUESTIONNAIRES DESIGNED BY THE TEACHER

Another way of using questionnaire surveys is for the teacher to design the questionnaire, but to involve students in collecting, representing and analysing data. This saves time and the questionnaires produced in this way are likely to be of higher quality than those produced by students. In Example 3 the questionnaire was designed by a teacher, Richard Davies, for year 8 students to investigate six hypotheses related to tourism patterns and preferences in Brentwood, Essex. The procedure for this enquiry is set out in Example 3a.

Example 3: Tourism survey. Source: Richard Davies.

(a) Procedure

The enquiry procedure was as follows:

Lesson 1

- The teacher introduced the focus of the enquiry
- Students worked in groups using thought showers on the project design
- Students identified questions, expected patterns and factors that might influence travel
- The teacher used the students' questions to develop a questionnaire using most of the questions identified by the students (Example 3b).

Lesson 2

Students wrote the first part of the report, supported by a task sheet (Example 3c) and a list of hypotheses, based on the discussion in lesson one (Example 3d).

Lesson 3

- Each student in the class used the questionnaire in Brentwood High Street to interview four people
- The teacher collected the questionnaires and entered the data into an *Excel* spreadsheet and prepared tables ready for students to use.

Lesson 4

Students analysed the data using *Excel* supported by a further task sheet (Example 3e).

Homework

Students completed their reports (Example 3).

Assessment

Teachers highlighted the parts of the level descriptions in the attainment target for geography, and applied these to the completed reports. Students were given level grades plus comments.

Example 3: Tourism survey. Source: Richard Davies ... continued

(b) Tourism questionnaire

Investigation into the patterns of tourism of the people of Brentwood

1. How many holidays do you go on every year on average?

   0   1   2   3   4   5+

2. Which country was the destination for your main holiday last year?

   UK France Spain Germany Italy Portugal USA Ireland Other ....

3. Where will be/was the destination for your main holiday this year?

   UK France Spain Germany Italy Portugal USA Ireland Other ....

4. In which month of the year do you usually take your main holiday?

   Jan Feb Mar Apr May Jun Jul Aug Sep Oct Nov Dec

5. Which three of the following reasons are the most important for you when you make your choice of holiday destination?

Climate/good weather	Distance to travel
Quality of accommodation	Quiet and peaceful
Local culture and traditions	Cost – value for money
Lots of sporting activities	Speaking a foreign language
Good food/wine eating out	Night life (clubs and bars)
Geography/landscape	Activities for children

   Other reasons not listed above

   ....

6. With whom do you normally go on your main holiday each year?

   Family   Partner   Friends   Other ....

7. Which forms of travel do you use for the main part of your journey to get to your holiday destination?

   Car   Plane   Ferry   Coach   Train   Other ....

8. Where would be your ideal holiday destination in the UK? ....

9. Where would be your ideal holiday destination in the world? ....

Thank you very much for your time in helping me with my project.

After the interview: male/female   age: 0-18 18-30 31-55 56+

(c) Tourism task sheet

Setting up the investigation

Task

- Set up a *Word* document file, within which you will create your report on the tourist patterns of people in Brentwood.
- Your first task is to write an **introduction**, a list of **hypotheses** and a **methodology**.

The introduction

On the first page of your project, explain:
- What the investigation is about
- How you are going to collect the information
- What you will do with this information
- What geographical information or knowledge you expect to gain.

Hypotheses

- A hypothesis is a statement that can be tested to see if it is *true* or *false*
- It can be a prediction of what we expect to find out. It should be based on commonsense and is usually backed up with sound geographical reasoning
- You will need to write out a series of hypothesis statements that reflect what you expect to find following the questionnaire survey into people's tourism patterns in Brentwood
- Look at the sample questionnaire. From this you should be able to make a prediction about the most common answers, e.g. Most people in Brentwood go to France for their holidays, or Most people in Brentwood go on holiday in August
- You might even try to connect groups of people to their holiday patterns. For example, Most people aged 18-30 choose holidays with good night life
- You should be able to develop *at least six* hypotheses
- Underneath each hypothesis type out a reason why you believe your statement to be true.

Methodology

- This section of your report explains how you intend to collect your information and then how you will use the information to answer your hypotheses
- Tips: you should mention the following:
  - Questionnaire survey – whole class involved to collect around 100 responses
  - Use of ICT – spreadsheets and databases to sort data and create statistics (averages, percentages), graphs and tables. The use of maps to illustrate popularity of destinations
  - If you can think of any criticisms or alternative ways to do this research you can also mention these in this section
- Do not forget to go to the file in the shared work area that has a copy of the questionnaire, you should copy and paste this into your document.

(d) Tourism hypotheses list

Tourism patterns and preferences of people in Brentwood

A **hypothesis** is a statement that we can test true or false. We write a hypothesis as a prediction of what we think we will find out. Then we do a survey/questionnaire to see if we are right.

Hypotheses we will test on our visit to Brentwood High Street:

1. Most people from Brentwood *prefer* to holiday somewhere hot when they go abroad.
2. Most people from Brentwood *prefer* to holiday by the sea.
3. Most people in Brentwood *have been* on holiday at least once in the last year.
4. Most people in Brentwood *have been* on holiday to a Mediterranean holiday resort this year.
5. Value for money is the most important factor when people choose a holiday.
6. Most people book their holidays through a travel agent.

(e) *Excel* instruction sheet

Writing up the investigation

You should already have prepared your **introduction**, **hypotheses** and **methodology**. Your next task is to produce the necessary graphs and tables needed to answer your hypotheses statements. This section is called the **results**.

Remember you will need at least **one graph** for each hypothesis statement. You should then be able to **interpret** this graph. What this means is that you can describe what the graph is showing and pick out any **patterns**. From the evidence, decide if you accept your hypothesis statement as being true or reject it for being false. This is your **analysis**.

Getting started

To access the results of the questionnaire survey, you need to follow this path:

Double click:
'Shared documents' – 'Applications' – 'Read applications' – 'Geography' – 'Tourism enquiry'

Then click and drag the *Excel* file 'survey data spreadsheet' into your 'my work' area.

Then double click the file to open it. You can then save this file and all the work you complete in this lesson under a different file name within your 'my work' area.

Producing graphs

The spreadsheet has two pages – one of survey data and the other of prepared tables to help you answer some of your hypotheses. Study these carefully and use the ones most appropriate to your hypotheses. You may have to create some extra tables of your own.

Once you have opened the spreadsheet you need to sort the data so that it is easy to calculate totals or work out percentages to put into the tables, for example, 'Most popular destination' choices last year.

On the data page click the menu 'data' and then 'sort'.

Select the first 'sort by' box and choose the column title 'Destination last year'.

Now the whole spreadsheet has been sorted to make it easier to count the number of people choosing each destination. Click between the data page and tables page to complete the tables. The first table has been completed for you.

On the data page click the menu 'data' and 'sort'.

Select the first 'sort by' box and choose the column title 'Gender'.

Then select 'then by' and choose the column 'Destination last year' and then 'ok'.

To produce a graph you highlight a completed table and follow the chart wizard instructions. Remember to produce a range of appropriate graphs for each hypothesis.

Example 3: Tourism survey. Source: Richard Davies ... continued

(f) Extracts from students' reports

Ashia

My hypothesis was: 'People aged 18-30 book holidays with a good nightlife'. The hypothesis was false. According to my graph 18-30 year olds prefer a holiday where there is hot weather. I found this out by sorting the age column and tabling the reasons what affects their choice of holiday.

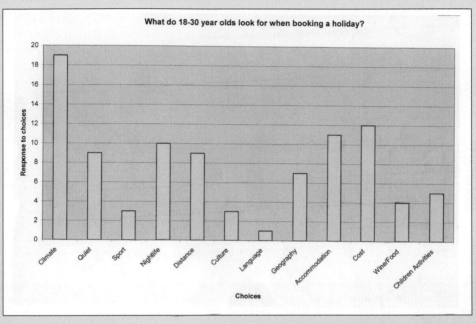

Conclusion

I have found out that you can expect a certain answer to a question but the answer that you get can be surprising. For example I would expect 18-30 year olds to go on holiday with their friends but according to my graph they prefer to go on holiday with their families.

I am pleased with the work. The only thing I would change is that I think I could have put more thought into the graphs that I have produced and I could have explained each graph in more detail. If we had more time to do the project it would have been better to ask more people so that we had more accurate results.

Sadia

Summary

I found out that people like to go to a hot sunny place by the coast about two-five times a year. When people book a holiday the three key things about booking a holiday are it has to be sunny, there has to be plenty to do, and the holiday has to be value for money. Most people go on holiday with their family and go in July and August.

Evaluation

When I was told that we were going to study tourism I thought it would be really boring but it wasn't. We had to design and make a questionnaire and then go and ask people on Brentwood High Street to answer it. I thought it would be really embarrassing but it wasn't. We only had time to ask one person, so if we did the project again I would have arranged to have more time in the High Street asking people to answer the questionnaires. This would have given us bigger and maybe better results. I think that next time I would do more research into tourism and how it destroys many parts of the world from litter to killing animals from noise and pollution.

Although the final choice of questions and hypotheses was the teacher's, the students were involved in the setting up of the project. Extracts from their work (Example 3f) reveal that they learnt a lot from the experience in terms of: knowledge; skills; awareness of some of the limitations of their survey in terms of accuracy and sample size; personal confidence in interviewing members of the public. They also enjoyed the experience.

Some of the extracts show that students would like to have used their own questions and explored different issues. Richard Davies, reflecting on this issue said:

'There is huge scope for differentiation here. Students could make their own questionnaires where ICT access allows and (print off copies for their class). But, some students will need support to develop questions and hypotheses and to identify patterns and factors. The use of writing frames is useful here. For example you could give some students part of a question and they could finish it, or an unfinished hypothesis for them to complete. Students can be prompted so that their hypotheses tie in with your expectations. You do, though, need to be adaptable to students who display initiative and originality.'

USING THE RESULTS OF QUESTIONNAIRE SURVEYS

Another way of introducing students to survey techniques is to present them with collected data (either real or imagined) and involve them in stages 6-8 of the stages identified in Figure 2 (page 156). This is much less involving for students, but can introduce them to different graphical techniques, give them choices in representing data and provide an opportunity for them to summarise findings and evaluate data. Example 4 shows a worksheet which enables students to develop the skills of analysing data collected through a survey. It has the advantages of presenting students with the raw data in tables, of

Example 4: Study of Padley Gorge, Derbyshire, within the Peak District National Park.

Finding out about Padley Woods

An A-level geography student has carried out a study at Padley Woods. He wanted to know how the area was used by visitors.

Using a questionnaire, he asked 100 people visiting the area a series of questions. This is a copy of some of the data that he collected.

1. Show the answers to the questionnaire using three different types of graph. Each must have a title and a key (if needed).

2. Describe what each graph tells you about visitors to Padley Woods.

	Car ₴₴₴ ₴₴₴ ₴₴₴ ₴₴₴ ₴₴₴ ₴₴₴ ₴₴₴ ₴₴₴ ₴₴₴ ₴₴₴ ₴₴₴ ₴₴₴ ₴₴₴ ₴₴₴ ₴₴₴ ₴₴₴ ₴₴₴	Bus ₴₴₴ ₴₴ ₴₴	Train ₴₴₴ ₴₴₴	other ₴₴₴
How did you get to Padley Woods today?				
How will you spend your time here?	Walking ₴₴₴ ₴₴₴ ₴₴₴ ₴₴₴ ₴₴₴ ₴₴₴ ₴₴₴ ₴₴₴ ₴₴₴ ₴₴₴ ₴₴₴	Picnicking ₴₴₴ ₴₴₴ ₴₴₴ ₴₴₴ ₴₴₴ ₴₴₴ ₴₴₴	Cycling or mountain biking ₴₴₴ ₴	other ₴₴₴ ₴₴₴ ₴₴₴ ₴₴
Do you think the area has been damaged by visitors in any of these ways?	Litter ₴₴₴ ₴₴₴ ₴₴₴ ₴₴₴ ₴₴₴ ₴₴₴ ₴₴₴ ₴	Footpath erosion ₴₴₴ ₴₴₴ ₴₴₴ ₴₴₴ ₴₴₴ ₴₴₴ ₴₴₴ ₴₴₴ ₴₴₴ ₴₴₴ ₴₴₴ ₴₴₴	Damaged verges from parked cars ₴₴₴ ₴₴₴ ₴₴₴ ₴₴₴ ₴₴	None of these ways. ₴₴₴ ₴₴₴ ₴₴₴ ₴₴₴ ₴₴

Write about: How people get to the area, How people spend their time, The damage which people have noticed.

3. This questionnaire was carried out during two sunny weekends in summer. Suggest how the answers might have differed if it had been carried out in winter?

4. Suggest one other question which could have been included in the questionnaire and explain why you think it would have been a useful question.

Figure 3: Literacy support for surveys.

Students who need additional support for written work could be provided with the following:

1. Geographical vocabulary related to what is being investigated.

2. Data-handling vocabulary relevant to the survey, e.g. line graph, mean, survey, questionnaire, interpret, bias, distribution, raw data (see Figure 7, Chapter 8, page 99).

3. A few phrases and sentence starters can be devised for a particular enquiry, for example:

- The aim of the survey was to find out …
- I expected to find that …
- The questionnaire asked people about …
- The data collected for question 1 is shown on a … graph
- The graph shows …
- On the one hand …
- On the other hand …
- The most important things I found out from the survey were …
- The most surprising thing I found out was …
- I did not expect to find …
- If I carried out the survey again I would …
- The survey might be misleading because …
- In conclusion …

providing choices in how to present the data, and of providing them with opportunities to summarise findings and evaluate the data. It should be emphasised that statistics represented on this worksheet were modified from data collected by year 12 students. Although the figures provided might be reasonable, they are not presented here as reliable data on the use of the Peak District National Park. Rather this worksheet is presented as an exemplar that could be adapted and applied to a wide variety of contexts.

LITERACY DEVELOPMENT

Literacy can be developed through survey by providing opportunities for students to:

- talk with each other about investigating an issue
- devise their own questions and construct questionnaires
- use vocabulary encouraged in the mathematics curriculum related to data handling
- communicate with adults other than teachers
- generalise about data collected
- use meaningful data for report writing

Some students may require additional support in the form of lists of vocabulary, connective words and sentence starters (Figure 3).

INTRODUCTION

We all know the world from what we learn directly through our experiences and from what we learn indirectly, e.g. from other people, the media and through formal education. This is our personal geography. Everyone's personal geography will be different, as we have had different direct and indirect experiences of the world within the cultures within which we live. Our personal geographies, which are constantly being reformed, are made up of mental maps, countless images and memories of places and situations, ideas of why things are the way they are and feelings about places and issues. Our personal geographies shape the way we see the world: they influence what we see, what we pay attention to, what we take for granted, and how we explain it to ourselves. What we remember tends to be selective in ways that support our pre-existing ways of seeing. Our personal geographies influence how we act within our environments and how we ourselves contribute to shaping the cultures within which we interact.

'To be human is to live in a world that is filled with significant places: to be human is to know your place'

(Relph, 1976, p. 1).

Throughout this book the importance of students relating their existing knowledge to new knowledge has been emphasised. This chapter adds more emphasis. It sees students' personal geographies as worthy of investigation in themselves.

The chapter starts by providing a rationale for giving personal geographies particular attention. It then identifies key questions for enquiry and suggests how they can be incorporated into the curriculum to provide a wide range of learning opportunities. Finally it suggests activities through which the personal geographies of students and other young people can be investigated.

RATIONALE

There are several reasons for focusing enquiry work on personal geographies. First, they are relevant to much of what is studied at key stage 3. Young people shop, go on holiday, move house, and experience the quality of life in the areas in which they live and places they visit. There is something students could say about all these things. They are not, however, always given the opportunity to do so. This is partly because in most geography classrooms students study people generally; they learn about where people shop, where people go on holiday, about people who migrate, and about the quality of life of people in different parts of towns. But this is all about other people, about people too generalised for many young people to relate to. It is also partly because most lesson plans do not provide space or time for personal anecdotes; students in secondary schools learn quickly that they are not expected to volunteer information from their own experience.

Second, this type of enquiry can help students make sense of the hotchpotch of images, memories, mental maps, ideas and feelings that constitute their own personal geographies. The impact of school geography will be greater if it helps students make sense of their own experiences of place and space in their lives outside school, as well as to make sense of what they learn in the geography classroom.

Third, geography in higher education is giving increasing attention to the geography of difference. Human geography is now concerned not only with generalisations but also with the ways individuals and groups experience space and place differently (Jackson, 2000). Generalised accounts can obscure the reality of the lives of individuals and groups and the selective ways in which they know the world. For instance, different groups of people have their own geographies of shopping. Shopping is a different experience for a 13-year-old girl, a 13-year-old boy, a mother with young children, a disabled person and a pensioner. Individuals go to different shops for different purposes and they have different views of shopping and shopping centres. Is shopping studied at key stage 3 as a differentiated activity or a generalised activity? Are students' experiences of shopping made relevant? Similarly, there are different personal geographies of travel and tourism. The concept of 'holiday' for a young British Muslim who has visited Pakistan, Mecca and many relations living in England is different from the concept of 'holiday' for students who have travelled little beyond their home town, and is different yet again for students who have experienced holidays in Britain and other parts of the world. Are travel and tourism studied at key stage 3 as differentiated or generalised experiences? Are students' experiences of travel made relevant?

There has also been an increase in the amount of research into the geographies of children and young people. Of all the new developments in academic geography, this is possibly of most interest to students. For example, Massey's work on youth cultures (1998) is highly relevant. She argues that all youth cultures are hybrid, the products of complex processes of interaction between the local and the global, involving 'active importation, adoption and adaptation' of 'contacts and influences drawn from a variety of places scattered, according to power-relations, fashion and habit, across many different parts of the globe' (Massey, 1998, p. 126). Massey provides a wonderful challenge for teachers developing geographical enquiry at key stage 3 when she writes:

'It would be a fascinating exercise to try to map even just a few elements of this complex of cultural influences and the different kinds of social relations and social power which they involve and express. A map of the world would certainly reveal some parts of the world as foci of more powerful influences than others. More modestly, the exercise might be tried for a particular youth culture, in order to capture the geography of influences (both inward and outward), their evolution over time, and the power relations which they embody' (Massey, 1998, p. 126).

What a wonderful focus for a class enquiry!

Another aspect of youth culture that could be studied in school is young people's use of space and how access to it is controlled for them and by them. Matthew *et al's* research (2000) emphasised the importance of public space, in particular the 'street', to young people. They found that different individuals and groups made different uses of the 'street'. For all young people, however, it represented a kind of 'thirdspace' in which they could develop their identities in spaces 'betwixt and between, neither entirely "owned" by young people nor fixed as adult domains' (Matthew *et al.*, 2000, p. 77).

KEY QUESTIONS

Key questions for enquiries focusing on personal geographies could include the following:

- What do I know from my own experience?
- Where have I been? Why have I been to these places? For what purposes?
- What are my attitudes towards the places I have experienced?
- In what ways am I aware of places through indirect experiences?
- What are my attitudes towards places I know of indirectly?
- How do I use public spaces?
- How is my use of public spaces controlled by me?
- How is my use of public spaces controlled by others and why?
- How do my experiences compare with those of other young people?
- How can we explain young people's experiences of place and space locally and in other places?

WHERE DOES PERSONAL GEOGRAPHY FIT IN?

Enquiries into personal geography and into the geography of other young people can be incorporated into many units of work in the key stage 3 geography curriculum, on both places and themes. Possibilities include:

- **Place**: students' direct knowledge and indirect knowledge of, e.g. the local area, the UK, the world
- **Place**: the ways in which students' lives are connected with other places, e.g. through relatives, friends, food, clothes, music, sport, possessions
- **Place**: how other young people experience the world, e.g. the lives of street children, refugees
- **Population**: migration, e.g. students' experiences of moving home; the experience of young refugees
- **Economic activities**: shopping, e.g. students' personal experiences and attitudes
- **Leisure activities**: students' and young people's experiences, e.g. use of space in the school and its grounds; use of public spaces: access and restrictions.

LEARNING OPPORTUNITIES

Enquiries focusing on personal geographies and the geography of young people provide opportunities for students to:

- value their existing knowledge and experiences
- make sense of their existing knowledge through relating it to broader geographical ideas
- become aware of the different ways in which their own lives are connected to other places in the world
- become aware that the world is experienced differently by different people and different societies
- compare and contrast their own experiences with those of other people, both within the school and those studied from data
- become aware of the way their own experiences shape their attitudes
- increase locational knowledge
- develop skills of:
  - o mapping
  - o recording
  - o communicating
  - o analysis.

ACTIVITIES

The following activities can be developed into complete enquiries to be combined with other enquiries to form a unit of work. Alternatively, activities focusing on personal geography and the geography of young people can be one part of an investigation planned to take place in one or more lessons. The activities are categorised under five headings:

- survey
- affective mapping
- mapping personal experience and knowledge
- relatives bingo
- the geographies of children and young people.

Example 1: Surveys of personal experience of place (year 7).

(a) Procedure

Key questions

- What places do I know from personal experience?
- Why have I been to these places?
- What places do I know about indirectly?
- How have I heard about these places?

Resources

Questionnaire surveys with sections for the local area (Example 1b), the UK and the world (similar questions in each section)

Maps of the local area, the UK and the world, for reference

An account of the experiences of year 7 students (Example 1c)

Starter: Listen and remember

- Teacher introduces students to the focus of the enquiry: individual's personal geographies.
- Teacher reads out an account of a year 7 students' personal geography.
- Students, in pairs:
  - list the kinds of places the student had visited
  - list the kinds of places he/she knew about but had not visited
  - hold a brief discussion: What kinds of places had he/she visited? Why had he/she visited these places? What kinds of places had he/she heard of? How had he/she heard of these places?
- Teacher introduces questionnaire

Using data: Creating data

Individual work: students complete the questionnaire, using the maps for reference.

Making sense of data

Extended writing: students write about their experiences of the local area and, if there is time, of the UK and the world. Some students might need some sentence starters and vocabulary.

Reflection on learning

- Sharing information: What kinds of places have you been to locally, UK and the world? Why did you go there? What have you learnt about the personal geographies of people in this class?
- Focusing on locational knowledge: on maps of local area, the UK and the world, finding places that people have visited.

(b) Survey sheet

The local area

1. Make a list of the places you have visited in .... (the local town).
2. Choose five of the places you have visited. In the table below, write the names of the places and give the reasons why you have visited them.

	Place	Why I have visited it
1		
2		
3		
4		
5		

3. Of the places you have visited in .... which is your favourite place and why?
4. Of the places you have visited in .... which place do you dislike most and why?
5. Make a list of places you have heard of in .... but have not visited.
6. Choose five of these places. In the table below, write the names of the places and how you have heard about them.

	Place	How I have heard about this place
1		
2		
3		
4		
5		

Note: Similar questions can be used in surveys of the UK and the world.

(c) Students' experience of place

The examples below and opposite are based on interviews carried out by PGCE students with year 7 students. The aim of the interviews was to try to find out students' direct and indirect experiences of Sheffield, Britain and the world. All names have been changed.

Shafeeq

Shafeeq has lived in Sheffield all his life. His family originated in Rawlpindi in Pakistan, His grandfather moved with his family to England. Shafeeq's father works in the grocery industry while his mother sells clothes in a local textile shop. His knowledge and understanding of the local area is very good. He knows several parks because he plays sports, especially cricket and football with family and friends. He visits Meadowhall for shopping, but was not old enough to go alone. He knows of other schools through people he was with at primary school.

He knows where to find both the football grounds in Sheffield, but he supports Manchester United. He knows about other sporting facilities including Ponds Forge (swimming) and Don Valley Stadium where a friend had won a sprint race.

He was very excited about a visit to London on a trip organised by his local Youth Club. The place Shafeeq knew most about was Warrington. He had visited his cousins there countless times. He likes Warrington because it is quieter than Sheffield and because there is one street in which almost everyone's family is of Pakistani origin. He also has family links with Birmingham and Walsall.

Shafeeq's only experience of another country is from visiting relations in Pakistan. He also knows about his father's visits to the Netherlands.

Most of Shafeeq's knowledge comes from friends and family and from his own individual experiences. He has gained much less knowledge from television or newspapers.

Saira

Saira lives a five-minute car journey from the school, with her 18-year-old brother and mother. Her mother came over to Britain from the Yemen at the age of 19 and settled in Sheffield with Saira's father. Her parents have since separated and Saira's father now lives and works in London. Saira has lived in Sheffield all her life and has widespread knowledge of the city. She enjoys going to Meadowhall where she goes shopping, goes to the cinema and the arcades. She also enjoys going into town with her mother. She has also visited Concord Park, the dry ski slope, the ice-skating rink, Ponds Forge swimming pool and Sheffield Arena where she enjoyed watching Destiny's Child in a concert. At weekends, Saira attends an Arabic School where she practises her Muslim religion. Her indirect knowledge of Sheffield is more limited. She has heard of the University of Sheffield and Sheffield Hallam University and Crystal Peaks (a shopping area), but does not know where they are.

Saira has visited many cities across Britain. She has visited friends and relations in Nottingham, Birmingham and Leeds and has been shopping in Liverpool and Doncaster with her mother. Saira has been to London at weekends to visit her father and stepmother. She has also been on family trips to Derbyshire where she enjoyed spending time in the countryside, having picnics and feeding the ducks. She likes Derbyshire because there is more open space compared to Sheffield. Saira's indirect knowledge of Britain includes theme parks such as Thorpe Park and American Adventure. She has heard of Scotland through the television but doesn't want to go there because she doesn't like the way they talk.

Much of Saira's direct experience of the world comes from holidays and visiting friends and family. She has visited an uncle in Egypt and enjoyed seeing the pyramids. She had also been to Yemen twice to see her mother's family and had been to Disneyland in Florida and Paris with her family.

Much of Saira's knowledge of places comes from her direct experience, mainly through visiting family and friends. Saira says that she enjoys living in Sheffield and would like to stay here when she grows up.

SURVEY

In this activity the data for the enquiry is provided by what the students know from direct or indirect experience. The data is collected through the use of a questionnaire survey to be completed by each student individually. In Example 1 a survey was used to find out year 7 students' experiences of Sheffield, the UK and the world. The focus of the survey was on what the students had experienced directly and what they had heard of indirectly. The questions were based on ideas

Example 1: Surveys of personal experience of place (year 7) ... continued

Sadie

Sadie is 12 years old. She does not like to walk through the park unless it is busy as she feels it is dangerous and this is why she catches the bus to school, which takes five minutes. She likes going to fast food outlets like Pizza Palace and others near her home. In Sheffield she visits Meadowhall, an enormous out of town shopping centre, which she goes to with friends on the bus. Sadie also visits Crystal Peaks (a smaller shopping centre) and the city centre. She knows Fargate and The Moor (shopping streets). She travels by bus and car to these places with friends or family.

Her father lives in Birmingham, which she goes to at the weekend and at holiday times. She goes by car and knows it takes a long time. She spends a good proportion of time in Birmingham. She has also been to Blackpool and has visited Whitby on a field trip.

Sadie has had a lot of world travel experience, visiting relatives and on holiday. Her parents were both originally from Jamaica and her knowledge of the world was better than that of the UK. She has visited Jamaica three times and has also been to Kenya and Ethiopia more than once as her mother has friends there. Sadie has been to Addis Ababa, but did not like it, as it was hot and dusty.

Lee

Lee is 12 years old and has lived in Sheffield all his life. He knows the local area well, because of journeys to and from school, shops, sports centres, friends' house and other similar activities. He has visited Meadowhall, the town centre, the Peace Gardens and the Odeon cinema with his mother and younger brother. He has also heard of a lot of places in Sheffield, such as Ponds Forge, the Crucible Theatre and Firth Park.

All his holidays have been in Britain with his family, most of the time camping. He has been to Wales, Scotland and Cornwall. He knows that all these places take a long time to reach by car from Sheffield. Of the places he has visited, Lee has positive views of Cornwall and Wales because it was good weather when he went, but he thinks that Scotland is not so nice because it rained a lot when he was there. He has heard of lots of others cities in Britain including Manchester, Liverpool, London, Newcastle, Birmingham, Bristol, Cardiff and Glasgow but was not sure where they all were.

Lee has not travelled abroad but has heard of a lot of countries, including India, China, Pakistan, Korea, Japan and Afghanistan, USA, Mexico, Brazil, Chile and Peru, New Zealand, Australia. He has gained some of this information from television and newspapers and some from school. Most of it has been gained from playing a 'spinning the globe' game with his brother and friend. They take it in turns to choose a country, then spin the globe and the other ones must find the country. He enjoys this game and likes finding out about the world. Lee really wants to go to Brazil because he thinks there would be lots of exciting things to do there and there would be rainforests. He did not want to go to America because he was frightened of all the guns.

from models devised by Goodey (1971) (Figure 1) and by Matthews (1992) (Figure 2). Both these models emphasise the role of both direct and indirect experience in young people's environmental knowledge. Goodey's model was constructed before the impact of information technology, a source of indirect knowledge for many young people. Students could be invited to modify Goodey's model to make it fit their experience. Matthews' model also includes the categories of direct and indirect knowledge, but in addition includes a 'lens of experience' through which experiences and messages are mediated. How they are mediated will be influenced by age, sex, ethnicity, class and cultural experience. Students would be able to discuss ways in which they think their own experiences might have been influenced by their age, gender or ethnicity.

Figure 1: Goodey's model of sources of environmental information. Source: Goodey, 1971.

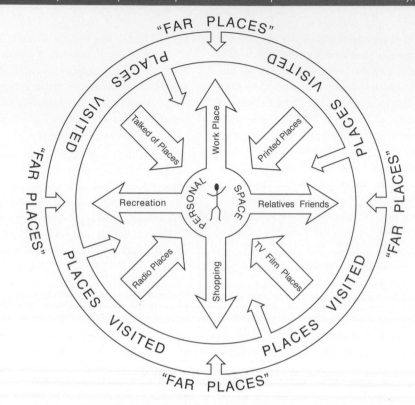

A survey with a different focus is shown in Example 2. In this survey year 8 students identified where they were allowed and were not allowed to go within their local area and within Sheffield, and what they felt about these places. The importance both of friends' homes and local outdoor areas became apparent. It was interesting that most students liked noisy places and disliked quiet places, in contrast with what is presented as environmentally preferable in many school textbooks which reflect general views, rather than the views of young people.

Figure 2: Matthews' model of children's environmental cognition. Source: Matthews, 1992.

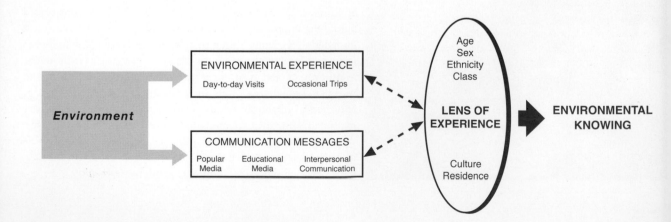

Example 2: Survey on personal geography. Source: Stephen Brady, Graham Fairclough, Helen Prescott and Susan Wells.

(a) Procedure
Key questions
- What are the best places to go with friends?
- What restrictions are there on year 8 students?
- What do year 8 students feel about the places they know?

Resources
Questionnaire (Example 2b)

Starter
- Teacher tells anecdotes about where he/she was and was not allowed to go on his/her own when he/she was 13 years old, and explains why. He/she tells the class about the kinds of places he/she liked and did not like.
- Teacher wonders what the experience of class is and introduces questionnaire.

Using data
Students answer questions from questionnaire on paper.

Making sense of data
- Students draw their own mental maps of the local area, marking on all the places they are allowed to go and indicating where they are not allowed to go.
- Students add symbols to indicate their feelings about places they have marked (see activity on affective mapping below).

Reflecting on learning
Sharing information through plenary debrief: Which kinds of places have you marked on your maps? Which kinds of places are you not allowed to go to on your own? Why not? What kinds of places are your favourite places? Why? Are there any places you do not like? Why not?

(b) Questionnaire survey of places visited and why
Think about the area around where you live:

1. Which are the best places to go with your friends? _____
   _____
2. Why? _____
3. Which places aren't you allowed to go to on your own? ___
   _____
4. Why? _____
5. Which of these places do you like?_____
   _____
6. Why? _____
7. Which of these places don't you like? _____
   _____
8. Why? _____

Think about other places you go to in Sheffield (places that are too far to get to by walking)

1. Which places do you go to in Sheffield? _____
   _____
2. Which of these places do you go to on your own or with friends?_____
3. Which places do you have to go to with someone older?___
   _____
4. How do you get to these places (e.g. tram, car, bus)? _____
   _____
5. Which of these places do you like?_____
   _____
6. Why? _____
7. Which of these places don't you like? _____
   _____
8. Why? _____

Example 3: Affective mapping of the school environment.

(a) Procedure
Key questions
- What are my feelings about different places in the school environment?
- What can be done to improve the school environment?

Resources
- Outline maps of the school and its grounds
- Students' personal knowledge

Starter
- Teacher explains that it is possible to map feelings.
- Teacher asks: What are your favourite places within the school and its grounds? Why? What are your least favourite places within the school and its grounds? Why?
- Teacher presents overhead transparency with a few symbols, e.g. happy, sad.
- Class discussion on other types of feelings about the school environment.

- Students devise symbols (adding to them as they are doing the map).

Using data
Students mark symbols on outline map, make a key and list reasons for feeling next to the key (Example 3b).

Making sense of data
- Students study their maps to see the overall pattern.
- They identify its key features.
- Plenary discussion of homework task: designing a poster or writing a letter to the head teacher about the school (Example 3c).

Reflection on learning
- Plenary discussion: sharing findings.
- Debriefing prompts: Did you experience any problems in mapping feelings? What patterns have been identified? How could the school and its environment be improved?

Example 3: *Affective mapping of the school environment ... continued*

(b) Daniel's map of feelings towards his school environment

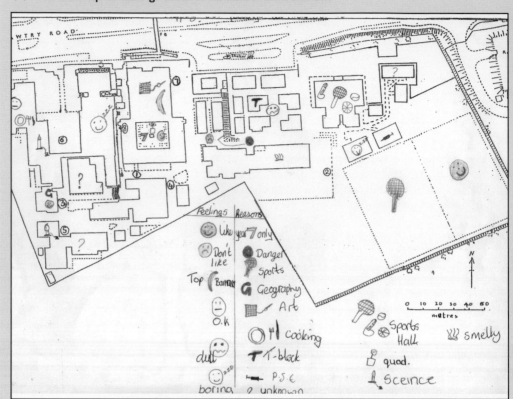

(c) Extracts from students' letters (reproduced as written) to the head teacher

Kyle

'I think some places are good but still need improving, and some places just need rebuilding. The quad is OK but needs improving by taking the litter out of the fountain, taking the weeds out and put different coloured flowers in to make it look nice. Also near the tennis courts there are small areas for all the school to play football and people end up getting hurt because its so small, and there is a risk of bullying.'

Ben

'Firstly let me complement you on your school. But I'm hear to tell you the bad points. Most of these issues will probebly be familiar. But I'm hear to stress them. My first issue is the Bus bey, it's smelly and there are cigeretts all aroud it.'

Shelley

'One more thin I find horrible is all of the graffiti in the little alcove at the side of the music block extremely awful. One thing I do like is the is the bright colours around parts of A Block, there is also some very good artwork decorating the corridors in the block which is in parts in one piece but in other people have destroyed it, I think that some broken pieces should be taken down because it ruins the rest of the brilliant decorative art.'

AFFECTIVE MAPPING

Affective mapping means plotting on maps the feelings that particular places evoke. Feelings are shown by symbols, possibly supplemented with annotation. Where patterns of feelings can be identified on maps, then areas can be shaded accordingly.

This activity has been used very successfully in the classroom. In Example 3, students mapped their feelings about the school grounds on an outline map. They used symbols they devised themselves and gave reasons for their feelings. They then wrote to the head teacher about their suggested improvements for the school (Example 3c). In Example 4, year 8 students constructed affective maps of their local areas, using their own hand-drawn mental maps as a base map. Students then applied these symbols to a list of places they visited in Sheffield.

In both these examples, students used symbols for:

- feelings, i.e. words that could be preceded by the words 'I feel … (e.g. happy, sad, bored)'
- descriptive reasons for feelings, e.g. colourful, smelly
- other information, e.g. places they could go to on their own.

Example 4: Affective mapping of the local area. Source: (a) Stephen Brady, Graham Fairclough, Helen Prescott and Susan Wells.

(a) Procedure
Key questions
- What do I know about my local area?
- What do I feel about it?
- Which places am I allowed to go to on my own?

Starter
- Teacher shows class an overhead transparency mental map they drew of their local area when they were in year 8 and identifies some features that were important to them.
- Discussion: What would students mark on their maps of the local area? What kinds of things would they mark on their own maps?
- Teacher puts overhead transparency overlay on mental map, on which are marked symbols of feelings about places.
- Discussion: (going through each of the symbols, including symbols for where allowed to go on own) Where in the local area do you have feelings like this?

Using data
- Students draw their own mental maps of their local area
- They make a key of different types of feelings
- They plot the symbols on the map

Making sense of data
- Plenary discussion: What kinds of places did you mark on your maps? Why?
- What kinds of problems did you have in drawing the map? Where are you allowed to go on your own? Where are you not allowed to go on your own? Why not? Have you added any other types of feelings you have?
- Individual activity: students write a list of at least five places they have visited in Sheffield and apply the same symbols as used on the map to indicate their feelings about these places (Example 4b). Each student then puts a sticker marking his/her places on a class map of Sheffield (colour coded for boys and girls).

Reflecting on learning
Plenary debriefing: What does the map of Sheffield show? How would you explain the pattern on this map? Are there any differences between girls' places and boys' places? What kinds of places are you allowed to go to on your own? Why? What kinds of places are you not allowed to go to on your own? Why? What are the main things you have learned from this enquiry?

(b) Applying values symbols to Sheffield

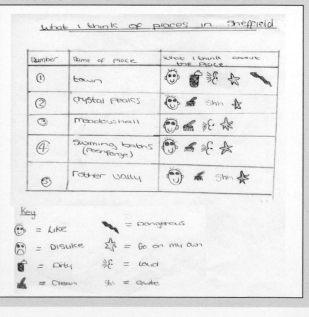

If the maps are supposed to indicate only feelings, then students will need some support in identifying feelings and developing the vocabulary to describe them. A good starting point might be the six categories of feelings identified by Charles Darwin in 1872:

- anger
- disgust
- fear
- surprise
- sadness
- enjoyment

If teachers want to develop this work further, they may want to expand this list. Recent work by psychologists have identified 412 separate feelings arranged in 24 groups! (Details of these can be found at: www.human-emotions.com).

An issue that has arisen in mapping feelings is that students may have conflicting feelings about the same place. For example, one student identified some cliffs in his local area as dangerous (fear) yet he identified this area as a place he liked. In another case, where students were drawing affective maps of where their relations lived, one year 7 student put both a smiley face and a sad face in mid-Wales. Her explanation was that she liked visiting her relations there when it was sunny because she liked the hills, but if it rained there was nothing to do.

Mapping personal experience and knowledge

If we want students to develop a good locational knowledge as a foundation for thinking about space and place, then a good place to start is with their own personal knowledge and experience. Mapping their own experience can encourage students to study atlases, to develop map skills and to develop awareness of locational relationships. Example 5 shows a map completed by a year 7 student. The task for the students was:

- to mark on the map, using guesswork, where the boundaries of England are
- then, using atlases, to correct the boundaries
- to name the four parts of the United Kingdom
- to mark on the map all the places they had visited
- to mark on the map all the places they had heard of

Depending on how well-travelled students in a class are, this activity could be extended to maps of Europe and of the world.

RELATIVES BINGO

The geography of family relations can be used as part of a study of place or population. A lively starter activity for such an enquiry is shown in Example 6, a lesson used in a year 7 class. A relatives bingo grid was used to collect personal knowledge which was then used as a basis for mapping and discussion.

Example 5: Mapping personal experiences and knowledge.

Example 6: Relatives bingo. Source: (a) Alison Munk.

(a) Procedure
Key question

Where do our relations live and why?

Resources

- Students' own knowledge
- Relatives bingo grid (Example 6b)
- Atlases
- Flip chart maps of the UK, Europe and the world (as appropriate for the class)
- Individual outline maps of the UK and the world.

Starter: Relatives bingo

- Students go round the class finding a different student for each box on the grid. When they have found an appropriate person, they write the name of the person and the relationship in the appropriate grid space
- When their grids are complete, students start the mapping activity.

Mapping activity

- Students use an atlas to find out the exact location of where their own relations live
- Students locate the places on individual outline maps (Example 6c) and on whole-class flip chart maps.

Plenary: Debriefing of relations bingo

What strategy did you use for completing the grid? (In a year 7 class one boy thought of the register and went round in alphabetical order, another peeked at someone else's grid, someone asked other students about incomplete boxes!) What types of relations did people mention?

Plenary: Discussion of maps

What patterns can you see? Where are most of the relations? Why do you think this is? Why have people moved to other places? (Students mentioned people moving to new jobs, families breaking up.)

(b) Relatives bingo grid

Find someone in the class who has relatives:

In Sheffield	In Europe	In a place with a hot climate	In the southern hemisphere
In a non-English speaking country	On an island	In a small village	Born in another country
Who have migrated from one country to another	Who all live in the same place	In a place he/she dislikes	In the countryside
In a place he/she likes	In a big city	Near mountains	In a continent other than Europe

(c) Example of a relatives world map

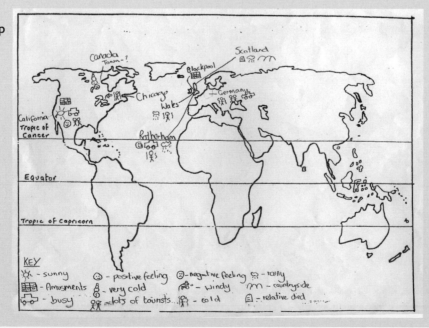

THE GEOGRAPHIES OF CHILDREN AND YOUNG PEOPLE

Students can increase their understanding of some geographical concepts and develop a wide range of skills as much through investigating the lives of children and young people as through investigating people generally. There are many possible ways of developing enquiry work based on the lives of young people, using the internet and information from organisations such as Save the Children and The Refugee Council. Example 7 was based on information found on the internet which was developed into an enquiry in which students investigated poverty in Nairobi through issues related to street children.

Example 7: Investigating street children in Nairobi. Source: (a) Rachel Atherton.

(a) Procedure

Lesson 1

Key question

What are some of the problems faced by street children living in Nairobi?

Resources

- Photograph of street children placed in 5 Ws question frame (see Nichols with Kinninment, 2001)
- Mystery statements (Example 7b), based on internet newspaper article (http://www.mg.co.za/mg/news/98apr1/15apr-kenya.html)
- List of suspects (Example 7c)
- Newspaper pro-forma with headline (completed Example 7d).

Starter

- Students provided with photograph and frame
- Students, in pairs, identify questions using the 5 Ws approach from Nichols with Kinniment's *More Thinking through Geography* (2001) asking questions beginning with What? Who? Where? When? and Why? Students write down preliminary ideas about the answers to their questions.

Mystery activity

- Students put into the role of investigative journalists
- Students provided with the newspaper pro-forma with given headline
- Students provided with mystery statements and a list of suspects
- Students work in groups sorting the statements in an attempt to solve the mystery.

Debriefing of mystery activity

Who do you think was responsible? (To different groups.) Why do you think this? Does this activity help answer any of the questions you identified at the start of the lesson? Which questions? Are there any unanswered questions? (Attempt to answer them.) What have you learnt from this? How reliable is the evidence? Do you think it would be possible to get a different view of these events?

Homework activity: Writing a newspaper article

What kinds of information do you think you need to include? (List on board.) What factual details might you include? If you had six paragraphs, how might you sort out the information? How could you start the article?

Lesson 2

Key question

What can be done to lessen the problems faced by street children in Nairobi?

Resources

Article about 'Life in Pendekezo Letu' (Example 7e) (based on information from http://www.childhopeuk.org/voices)

Starter

- Initial reading of the article
- Students in pairs try to correct the English in the article (supporting their own literacy development).

Using data: Analysis and reconstruction DART activity

- Students analyse the article by underlining negative comments in one colour and positive comments in another colour
- Students reconstruct the information by categorising it into comments about the street and comments about Pendekezo Letu.

Debriefing of DARTs activity

What negative comments did you underline? What positive comments did you underline? What did you put under the heading 'street'? Were these positive or negative? What did you put under the heading 'Pendekezo Letu'? Were these positive or negative? What do you think Pendekezo is?

Making sense of data: Writing

- Writing task: Write a short speech for a comic relief celebrity to make during video footage of Pendekezo Letu. The speech is a plea to the public for donations (Example 6f)
- Briefing: What facts are you going to include? How could you persuade people that they should donate some money? What would you want to be shown on the video footage?

Example 7: Investigating street children in Nairobi ... continued

(b) Street children: texts for mystery cards

1. After heavy rain, gutters overflow and sometimes street children drown in the filthy sewage water	15. Most migrants end up living in shanty towns on the edge of large cities.
2. Most days Moses found work carrying boxes for shoppers, begging or stealing.	16. Security guards avenged the death of their colleague by attacking street children sleeping on the pavements and in rubbish dumps.
3. Nairobi's police force is poorly paid and demoralised.	17. One security guard was stoned to death by a group of 100 street kids.
4. The rise in crime in Nairobi has led to many property owners employing private security guards.	18. People in shanty towns live in poor housing, have little or no access to education, clean water or medical facilities.
5. Some witnesses say that one of the security companies was responsible for the hit-and-run.	19. The government wants to round up street children to reunite them with their families.
6. Over 65% of Kenya's population live below the poverty line.	20. The number of street children in Nairobi has leapt to about 60,000.
7. Research has found that 6 in every 10 boys living on the streets have health problems associated with taking drugs.	21. Moses Mwangi was found dead on the night of the riot.
8. Police are finding it hard to cope with so many riots. They rely on many private security guards to help them out.	22. Moses had been sniffing glue to stop him from feeling hungry.
9. Security guards are armed with machetes, whips and guns.	23. In rural areas there are few jobs other than farming so people migrate in search of employment and a better quality of life.
10. Street children are often beaten by police when they are arrested.	24. The Mwangis moved to Nairobi to escape poverty in the countryside.
11. Street children are exposed to harassment and physical/sexual abuse. They lack food, water and sleeping space.	25. Moses was sent to the streets by his family to earn money. He was told not to return until he had.
12. Children injured in the riot made their way to hospital on foot or were carried by friends.	26. Mathare is Nairobi's largest and poorest slum.
13. A nurse from Kenyatta hospital said that many of the children sniffed glue and drank home-made spirits as they lay in casualty.	27. Moses' family live in Mathare.
14. The government is finding it difficult to keep up with the demand for services in cities because the population is increasing so rapidly.	28. Homelessness is a crime in Kenya.

(b) Street children: list of suspects

The suspects ...

Kenyan government   Kenyan police   Private security guards   Moses' family   Moses Mwangi   ?

Example 7: Investigating street children in Nairobi ... continued

(d) Lucy's completed newspaper pro-forma

Daily News   April 15 1998

Hit-and-run sparks riot by Kenyan street kids

A good summa of some of the issues. Who do you think is to blame for the problem? ✓

A battle between street children and private security guards in Nairobi left two people dead and 63 needing hospital treatment at the weekend. LUCY HANNAN reports.

here in Nairobi it is illegal to live on the street but people like Moses' (shown in picture) have little choice.

Moses was murdered after a raging battle between street children and security guards who wanted to avenge a fellow colleagues death after being stoned to death by 100 street children.

Here in Nairobi there are over 60,000 street children and over 65% of the Kenyan population live below the poverty line.

We interviewed one street child to see what he had to say about the government he said.

"The police can't cope they rely on the security guards who are privately payed or are armed with machetes, whips and guns, the government can't cope with the quickly increasing population."

So who's to blame for these tragic deaths I think more questions will/should be asked about this tragic event.

By Lucy

There has been much tension in Nairobi following the brutal murder of a security guard, security guards have been employed due to the incompetence of the police.

"Having no trust in the police isn't my personal feeling, everyone in the shopping community has it so we employ security guards." This is the view of many shop owners and it is a well known fact here that even if the police capture criminals, they are often beaten up before they reach the police station. I for one think this is disgraceful and is the governments fault

(e) DART activity: Pendekezo Letu

Life in Pendekezo Letu (Pendekezo Letu means 'our choice' in Kishwahili)

This extract was written about Pendekezo Letu by Perisi, a street girl:

'Life in Pendekezo Letu is more better that life on the Street of Nairobi. This is because on the street we are beaten by police men and the big boys we sleep without food or sometimes eat dirty food. Lack shelter and madicino when we fall sick and nobody cares about us at all.

In Pendekezo Letu we eat a lot of food we have people who care and love us we go to class and we are taught different subjects like English, maths, science, Kishwahili and above all we are given clothes and hope for the future.

In Pendekezo Letu we are involved in many activities like cooking sweeping washing clothes and dancing singing and sports.

We thank Pendekezo Letu for the support and good things it is giving to us and many other street children we pray that it continues with the good work it is doing for the needy and desperate children.

Thank you long live.'

Task

Underline all the negative comments in one colour and all the positive comments in another. Sort these into the table below:

The street	Pendekezo Letu

(f) Martha's writing based on Pendekezo DART

Script written by student for a charity video (reproduced as written).

'Before Pendekezo Letu the children lived on the streets. On the streets the young children were often beaten. They had a lack of food or even if they had any it was dirty food that we wouldn't eat. They had no medicine or shelter. Nobody cared about, the children felt very unloved. Pendekezo Letu offers the girls plenty of food, someone to care and love them. They get an education that gives hope for the future. They are given clothes as well. All the children are involved in activities such as cooking, sweeping, washing clothes and singing and dancing and other sports. Pendekezo Letu is away from the city so that the children don't go back to Niarobi's street. The aim of the organisation is to get the girls back to their families in a better state that were found. There are different organisations for boys as well as girls. This place needs our support.'

INTRODUCTION

We get to know the world mainly through how it is represented to us. Secondary school students will all have some direct experiences to bring to their understanding of geography. Mostly, however, they will learn about the world through how it is represented; by parents and friends; on television, including soap operas; on computer games; in films; in newspapers and magazines; in advertisements; and in geography textbooks. These representations shape what students know about other places.

What we know about places can be stereotypical and misleading. As part of a project linking the two countries a New Zealand child wrote about Britain in an e-mail: 'The houses are old and of an older style than seen in New Zealand. They often have two storeys and no front or back gardens like Coronation Street' (Valentine, 2001, p. 299).

According to Malcolm McLaren, the pop impressario:

'Being British is about singing Karaoke in bars, eating Chinese noodles and Japanese sushi, drinking French wine, wearing Prada and Nike, dancing to Italian house music, listening to Cher, using an Apple Mac, holidaying in Florida and Ibiza and buying a house in Spain. Shepherd's pie and going on holiday in Hastings went out about 50 years ago' (Poole et al., 1999).

Whereas John Major's image of Britain is rather different:

'Long shadows on county cricket grounds, warm beer, invincible green suburbs, dog lovers and, as George Orwell said, old maids bicycling to Holy Communion through the morning mist' (Poole et al., 1999).

It is worth developing enquiry work on representation of place not only because of misleading images, but also because of the growing importance of the marketing of place. Places are increasingly presented as commodities, represented by slogans and images. For example, places in the UK which have suffered economic decline suffer crises of identity and need to create a new image, partly for self-esteem and partly for marketing purposes. Massey (1996) noted 'a kind of schizophrenia' in local government agencies in areas of economic decline. On the one hand they needed to create positive images to attract new investment while on the other hand they needed to present negative aspects of the place to attract special funding. She presented newspaper headlines that illustrated these conflicts and the contrast between outsider images of Liverpool as a problem city and insider images of a much-loved city:

- No end in sight to Liverpool's despair
- Self-pity city
- Reborn city of vision and values
- Liverpool steps up to blow its own trumpet

Spooner noted the attention given to slogans in attempting to change the image of a city. Hull, in its search for a new identity, changed its slogan from 'Hull: dynamic European maritime city' to 'Kingston upon Hull: the Pioneering City' (2001, p. 293).

The images that are created about places through the use of slogans and photographs are created by particular people for particular purposes. Often the people who live in a place feel that it has been unfaithfully represented in such packaging.

Images of place are also created by news reports, films, soap operas or other television series. Again, local people often feel the images created are misleading. For example, many people in Sheffield felt that although the film *The Full Monty* (Cattaneo, 1997) publicised the city and even attracted tourists for a while, it represented the city in a negative way. Such images are pervasive; in recent years when I have mentioned to people from Korea, China, Germany, Italy, the USA, etc., that I live in Sheffield the first thing they have said is, 'Oh yes, *The Full Monty*'! We know places through their representations, through publicity for news events happening in these places, through their being chosen for venues for

'No representation of the world can be "neutral". Each representation necessarily has a particular perspective'

(Massey, 1995, p. 34).

filming, and through their marketing. When we study places we can study them through how they are represented. Geographical enquiry should include these wider representations of place in the data provided for study.

KEY QUESTIONS

Geographical enquiry work focusing on representation of place would answer questions such as:

- How has this place been represented?
- Who has represented this place?
- What does the representation of the place mean?
- Why has it been represented in this way?
- To what extent is this a fair representation? For whom is it fair?
- Why might a place want to market itself?
- How can a place market itself?

LEARNING OPPORTUNITIES

Representation enquiries can give students opportunities to:

- become more aware of their own images of place
- become more aware of how places are represented
- understand that secondary data about places are selections of reality
- analyse how places are represented
- create images for places for particular purposes

APPLYING THE KEY QUESTIONS

Representation enquiries are particularly appropriate for aspects of the key stage 3 geography national curriculum which involve the study of particular places. They could include:

- Study of a less economically developed country (LEDC) and of a more economically developed country (MEDC): e.g. What images do we have of this country? Where do they come from? How is this country represented in geography textbooks, in the media, in travel brochures?
- Settlement: What images can be created of the school/local area to give it a new identity? How can these places represent themselves to attract new projects or to attract funding? In what ways are other countries represented in the local region?
- Economic development: Where do industries want to locate their factories/offices (e.g. the car industry)? What kinds of places are suitable for major events or projects (e.g. Olympic Games)? How can places market themselves to attract these industries or events or projects?
- Tourism: How are places represented in tourist brochures? What images are not represented?

ACTIVITIES

There is a range of activities that can be incorporated into geographical enquiry focused on representation of place:

- contrasting images
- picture editor
- A fair view?
- marketing a place
- competing places
- representations of other places in the local region

The activities can be incorporated into a procedure that encompasses all the essential aspects of the framework for geographical enquiry (Figure 1).

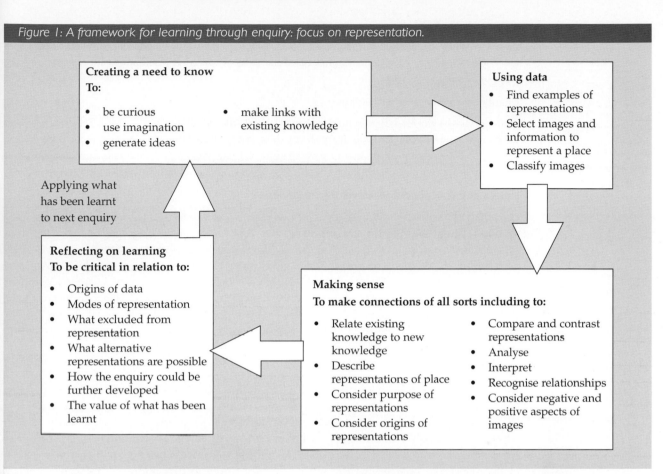

Figure 1: A framework for learning through enquiry: focus on representation.

CONTRASTING IMAGES

The key questions for the enquiry are: How is this place represented? Why is it represented like this? What other representations of this place are possible? In this activity the main task for students is to analyse an initial set of data providing a representation of a place and to compare it with a second set of data providing contrasting information (Example 1). Of course, neither set of data can portray what a place is 'really' like, as they will both inevitably be a selection of some sort. This activity draws attention to the selectivity of all data. It can focus on textbook images, tourist brochures, charity advertisements, slides or photographs of a country, advertisements and other representations on television.

Textbook images

As part of the study of a particular country, students analyse the textbook photographs chosen to represent the country. Categories for analysis could include: physical/human; rural/urban; more developed/less developed; what people (men, women, children) are doing in the photographs. Students discuss and generalise about the images of that country created in the textbook, discussing what is shown and speculating on what is not shown. Students then study a different representation of that country, for example, a few selected photographs or the front page of a newspaper from that country (via the internet).

Tourist brochures and holiday programmes

Students analyse extracts from tourist brochures or holiday programmes, using categories of analysis such as natural/created attractions of the place and types of activities for tourists and how the local culture is represented. They discuss what image the tourist brochure or holiday programme creates of the place and speculate on what might be excluded from the image. Students either search for, or are presented with, additional information, e.g. about how local people live, about less attractive features of the tourist destination. Students discuss the differences between representations for tourists and the other information.

Charity advertisements and videos

Students analyse the images presented in charity appeal advertisements or in a video. They discuss what the images suggest about this country and why these images have been chosen. Are people presented as passive and dependent or as able to participate in helping themselves? Students speculate on what images of that country have not been chosen. They then study contrasting images of the country.

Images of less economically developed countries

Hopkirk (1998) used contrasting slides of India to challenge perceptions of less economically developed countries. Before showing the slides, students had drawn two spider diagrams, one showing what students associated with the more economically developed world and the other showing what they associated with the less economically developed world. Hopkirk then used the Chinese Whispers strategy twice (Example 1, Chapter 9, page 115), first to show a slide presenting an image of a very poor part of India and then to show a slide presenting a smart modern part of an Indian city. Invariably the students imagined that the first slide was taken in a LEDC and the second slide in a MEDC and the way they portrayed the slides to those who had not seen it conveyed this strongly. Following this experience, students returned to their spider diagrams constructed from their pre-conceptions.

Advertisements

Advertisements can provide starting points for enquiries into representation of place. Jackson (2002) examined how contrasting television advertisements represented 'Indian' food in Britain, one focusing on India and difference and the other focusing on 'ethnic food' generally. A close study of such advertisements would enable students to understand how our images of place and the way we relate to other places are being shaped by the media. The study of food advertisements could lead on to a study of the connections between the UK and the country whose food is portrayed, through migration, employment, labour and food.

Representations of place on television

Enquiry work based on representations of place on television can link school geography to the world students encounter outside the classroom and can make students more aware of how our knowledge of place is shaped. The introductory sequences of BBC and ITV Football World Cup programmes in 2002, for example, provided interesting but contrasting representations of Japan and Korea. The ITV sequence showed flags of both countries, followed by traditional images of Japan: a drummer, a couple of geisha, and the rising sun, while the BBC sequence included urban shots of Tokyo and Seoul, the bullet train and neon lights, as well as traditional images. Korea was less represented in either set of images. Such sequences provide excellent data for analysis in the classroom, or as an introduction to a creative activity in which students are invited to make their own selection of images for introductory sequences for similar programmes.

PICTURE EDITOR

This activity needs to be set in a context in which one or more pictures needs to be chosen for a particular purpose. The task for the student is to study a range of images and choose one or more images appropriate for the purpose. The contexts in which images need to be chosen could include:

- A geography magazine article aimed at a particular age group, e.g. 8-10 year olds; 11-14 year olds. There is only space for two photographs: which should they be?
- A postcard to be sent to a friend or someone else to present a particular representation of a place, possibly using a postcard format from *Microsoft Publisher*, with students cutting and pasting selected images. Example 1b shows a postcard of Antarctica created by a student. Although the visit to Antarctica was imagined, the audience for the postcard was in fact real. The postcards were written following the visit by an Antarctic explorer to the class who answered questions they had devised, and were sent to him.
- The cover of a tourist brochure to attract family tourists (or any other categories) to a particular place.
- A leaflet promoting a town or area for a new project.
- A newspaper report. In Example 1 in Chapter 5 (page 56) students selected images of the flood disaster in Mozambique for a news report.

Example 1: Contrasting images: What is the place really like?

(a) Procedure

Opener

Students are introduced to the idea of representation. What images might be presented of this school in a brochure? What might be left out of a school brochure?

Students are introduced to the name of the place to be studied and where it is.

Activities

Group work: analysis of data

- Students are presented with the first set of data representing the place
- They make a list of all the things that this information tells them about the place
- They classify this information into different categories of information. The categories can be provided by the teacher, or devised by the students
- Interim whole-class debrief

- How is this place represented in the data? What different kinds of information are there? Why is it presented in this way? Who is the information produced for?
- Are there any ways in which you think this representation might be misleading? What do you think has not been shown?

Group work: search for conflicting data

- Students study data provided, or they search libraries, geography textbooks, the internet, etc., to find contrasting images of the same place. They select three contrasting images and summarise them.

Summary

Whole-class debrief

What new images have you found? In what ways are these different from the first set of images? Why were they not included in the first set of images? Are the images in the first set fair? Would any set of images be fair?

1(b): Olivia's Antarctica postcard, produced following the Picture Editor activity. Photos: Anna Gunby.

Dear Mr Lowe
The weather is very cold here in Antarctica but it is very sunny. I have not seen any penguins yet, but there are lots of birds nearer the sea. I have got very warm clothing to keep me warm and to stop me from getting frostbite! I have had hardly any sleep because the sun is shining all the time so you don't really know when it's night-time, but that's ok because I am never tired. All around me there is snow and ice. It is dinner time now I am having chicken soup and bread (that's if it's not frozen!) We are climbing tomorrow. Hope you are well back in England.
From Olivia

MR LOWE
JAGGED GLOBE
45 MOWBRAY STREET
SHEFFIELD
S3 8EN

The *Google* website has an excellent collection of images on a vast array of places and topics. These can be accessed from the 'images' banner at the top of the *Google* home page (http://www.google.co.uk). Because of the huge number of images on most places, teachers might prefer to pre-select images, and to make them available to students in a folder on an Intranet system.

A FAIR VIEW?

In this activity students select places to be visited by visitors in order to provide them with a 'fair view' of a town, a region or a country. Example 2 describes an activity used with year 7 students in which they planned a four-day 'fair view' tour of Britain for visitors from overseas. It would be interesting to compare these students' choices with those of other students from different parts of the country. What indeed is a fair view?

Example 2: A 'fair view' of England.

(a) Procedure

Key questions

- What do we mean by England?
- Where do you like to go in England?
- Where would other people like to go in England?
- Which four places would you choose to present a 'fair view' of England for tourists? Why?
- What provisions are there for tourists in these places?

Resources

- Outline maps of the British Isles, without boundaries marked on
- Worksheet: A fair view of England? (Example 2b)
- Note-taking frame for internet research (Example 2c)
- Letter (Example 2d)

Lesson 1

Starter

- Quiz: Where in the world could this country be?

Locational activity

- Students 'guess' where the borders of England are and mark them in pencil on the outline map.
- They mark on the map, in pencil, places they would like to visit (four places they have visited and four places they would like to visit), estimating or guessing their location.
- They check the location of the border of England (and correct if necessary), and name Wales, Scotland, Northern Ireland and the Republic of Ireland.
- Students check that the locations of the places they have named and marked and correct.

Where might different groups of people like to go?

- Brief discussion of worksheet (Example 2b) asking for some suggestions in each category
- Students complete the worksheet

- Feedback of students' ideas and discussion

What would make a fair view for tourists from overseas?

- Hand out letter (Example 2d) and read it
- Thought shower: What would be a 'fair view' for these tourists? What should be included?

Homework

- Think of four places for the visitors to see (Example 2e).

Lesson 2

Starter

- What did the class decide would be a 'fair view' of England?
- Introduce Internet research and the note-taking frame (Example 2c).
- Ensure that all students understand what they have to do.

Internet research (50 minutes)

- collecting information and at least one image of each place
- recording information on note-taking frame

Plenary

What have they found out? What images did they choose?

Homework

Search for images of the four chosen places.

Lesson 3

Research on the computers continues.

Lesson 4

In this lesson students plan the itinerary for the tourists, decide on best routes, modes of transport and how long each journey would take.

Homework

Students use their findings to produce a brochure which will become part of a classroom display.

Example 2: A 'fair view' of England ... continued

(b) Worksheet

A fair view of England?

Write in the speech bubbles what you think the person/people would like to see when visiting England.

Example 2: A 'fair view' of England ... continued

(c) Note-taking frame

Exploring England internet research

Aim: to plan a 5-day tour of England using information from the internet

Find out about	Place 1	Place 2	Place 3	Place 4
Tourist attractions				
Accommodation				
Events				
Anything else?				
Notes:				

(d) Letter

Dear ....

I am writing to ask you if you could arrange an itinerary for me and my friends for when we visit England in November. We will stay in England for five days, so can only visit four places. We want to see a fair view of England!

I look forward to hearing from you and seeing what you have planned.

Thank you for your help.

Yours sincerely,

A 'fair view' tourist

(e) Examples of places chosen

The places chosen by year 7 students in a Chesterfield school to represent a 'fair view' of England for overseas tourists included:

- Alton Towers (very popular choice)
- Blackpool
- Chesterfield
- Derby
- Great Yarmouth
- Lake District
- London
- Matlock
- Old Trafford
- Peak District (very popular choice)
- Skegness
- Sheffield

(f) Jeanette's 'Fair view' (as written)

A four day view of England

Hello and welcome to a four day view of England. You will be visiting Liverpool, Flamborough, Bridlington and finally London. Also you will visit attractions and features at each location. You'll visit the Royal Liver Building in Liverpool as well as the Albert Dock. Flamborough has its beachs lighthouse and you can also enjoy a days shopping. In Bridlington you will walk along the sea front and the harbour. Then eat in a posh restaurant. Finally to end your four day view of England you will go to the big city of London. Here you will visit the tower of London plus the London Bridge and take a stroll along the river Thames.

The activity can be varied in several ways. Instead of planning for visitors generally, students could plan for people their own age or for families where different interests could influence the choice of 'fair view'. Instead of planning a 'fair view' students could be asked to plan a 'horror tour' which represented what they thought were worst aspects of a place, or a 'super tour' which represented what they thought were the best aspects of a place. Students could be given a choice of these three options (fair tour, horror tour and super tour) so that their choices for representing the same place could be compared.

MARKETING A PLACE

In this strategy, students are presented with a purposeful context in which a place needs to market itself. Purposeful contexts could include:

- a bid by an area in economic decline to attract special funding, e.g. from the European Union (EU)
- a bid to attract a particular industry, e.g. a car factory
- a bid to attract a particular event, e.g. Olympic Games
- a bid to attract tourists to the area.

This strategy works best if care is taken to include the four essential aspects of geographical enquiry:

1. creating a need to know: discussion of the purposeful context in which the enquiry is set and what the bid might need to include
2. using data: collecting information relevant to the purposeful context, including: using existing knowledge, fieldwork experience if possible, and secondary information
3. making sense of data: (a) sharing information and discussing: what images might be selected and what might be excluded; ideas for possible slogans, and (b) recording these ways of representing the place in some way. This could include the use of a digital camera, a camcorder, or *Microsoft Publisher*, and could result in the production of photo packs, video or display materials
4. reflecting on learning: How do they represent the place? What has been selected? What has been excluded? To what extent are the images fair? Would any people living in this place object to these images?

COMPETING PLACES

For this strategy, students study or create representations of places competing against each other for the location of an industry or project or an award. The task for students would be to create a leaflet or to represent the place in a role play. The situation could be an imagined situation, but many real situations can be used, for example:

- Which of the six competing UK cities should become EU city of culture in 2008?
- Which one of competing UK towns should be given city status?
- In which country/countries should the 2010 World Cup take place?
- In which country should the 2012 Olympic Games take place?
- In which part of the UK should a specified industry locate a new factory?

Example 3 provides a possible procedure for an enquiry involving competing places.

Example 3: Competing places.

Procedure

Key question

Which city should be selected for EU city of culture for 2008?

Resources needed

Information, or access to information on some of the six cities in the long list: Bristol; Birmingham; Cardiff; Liverpool; Newcastle and Gateshead (joint bid); Oxford.

Starter

- Introduction to the competition to become the EU city of culture.
- Why might cities want to become the EU city of culture? What kinds of things could they use to promote themselves? Using a thought shower to record what kinds of features a city might use to promote itself.
- What criteria could be used for judging who wins? (The class could devise the criteria to evaluate the presentations.)
- Set up the groups.
- Give advance notice of the final form of presentation: role play.

Groupwork activity

- Students work in groups of 4 or 5, each group investigating the advantages of their particular city using data on the internet, or information provided by the teacher, or data selected by the teacher put into folders on the school intranet. They could be provided with both positive and negative information.
- Group work preparation for a 1-2 minute presentation at a role-play public meeting. Each group could be expected to locate the city, make a given number of key points and to present one visual image of their city.

Role play

- Students locate their city and present their key points and their selected visual image at a public meeting.
- They answer questions about their city.
- A decision is made by the teacher or by a group of students.

Plenary debriefing

The role play is debriefed. Which cities were able to make the best representations? What information was not presented? Which cities were the most difficult to represent and why? How does the decision made in the role-play compare with the actual decision?

Alternative forms of representation

Instead of representing the city in a role play, each group could present its city in a poster display or a display of small booklets. The criteria for judging the display could be determined at the outset and the decision on which city was chosen would be based on these criteria.

REPRESENTATIONS OF OTHER PLACES IN THE LOCAL REGION

The key questions for this investigation would be: What are the effects of the country being studied (e.g. Italy) on the local region? What are the most important effects? What do we gain from our links with Italy? What difference does Italy make to our lives? What difference do these connections make to Italy? What image do we get of Italy from these connections? This activity is appropriate when studying countries which have strong representation in the UK, e.g. Italy, USA.

The procedure could be as follows:

- Students use a thought shower to generate initial ideas on how the country is represented in the local region, or within the UK.
- These are categorised. Likely categories for Italy would include: food products, cafés and restaurants, commodities such as cars and clothes, advertisements for holidays, Italian people living or working in the area (e.g. footballers), newspaper reports, television programmes, advertisements.
- Students could investigate this further through a questionnaire survey, through mapping Italian restaurants, through studying newspapers, through mapping where Italian footballers play and mapping where they come from in Italy, through mapping places of origin of Italian food and products.

CHAPTER

15

FUTURES

'People have always wanted to get a preview, an early glimpse, of what the future holds. All of us – individuals, organisations, corporations, authorities or ministries – make plans and anticipate the future in our daily lives. In that sense we are all "futurists"'

(PIU, 2001, p. 2).

INTRODUCTION

This chapter focuses on futures enquiries, because of their particular relevance to secondary school students and because of the general importance of futures thinking in the world today.

The lives of students presently at key stage 3 can be expected to last well into the later years of this century. Research by Hicks and Holden (1995) shows that the majority of students of secondary school age think frequently about their future lives and the future of the world and that they respond with interest to activities discussing the future of their local area.

Another reason for focusing on the future is because of the increasing importance of 'futures work' within institutions, industrial organisations, local and national government and within international organisations. Thinking about the future is used as a creative process which can increase the range of policy options, can guide decision making and can identify opportunities and threats. This chapter draws on some of the key methodologies used in futures work (PIU, 2001).

KEY QUESTIONS: THE THREE PS

Futures work tends to focus on the three Ps – possible, probable and preferable – and their associated questions:

- **Possible futures:** What may happen?

- **Probable futures:** What is most likely to happen?

- **Preferable futures:** What would we prefer to happen?

How these questions are developed depends on what methods are used in futures thinking, quantitative or qualitative. Quantitative methods are concerned with forecasting from present data. Key enquiry questions related to forecasting could include:

- If these trends continue, what will happen in the short term (1-5 years)?

- If these trends continue, what will happen in the longer term (10 years or more)?

- How likely are these forecasts? Why might they be wrong?

- What are the likely consequences of these forecasts?

Qualitative methods are more concerned with feelings about the future, about attitudes, values and moral judgements. Key enquiry questions could include:

- What alternatives are there for the future?

- What would be the preferred future? (Hoped for futures, utopian futures, most just futures.)

- What would be the worst future? (Feared futures, disaster futures.)

- Who wins and who loses in what is envisaged for the future?

- What would I prefer?

Most futures thinking in institutions and organisations uses a combination of quantitative and qualitative approaches.

LEARNING OPPORTUNITIES

Futures enquiries can give students opportunities to:

- increase their knowledge and understanding of current trends

- increase their knowledge and understanding of processes of change

- consider the implications of change for themselves and for others
- think creatively about the future
- consider alternatives
- examine their own attitudes and values about the future
- become aware of opportunities people have to influence change (related to the citizenship curriculum)
- develop numeracy skills
- develop communication skills.

APPLYING THE KEY QUESTIONS

Much of the research into young people's thinking about the future has been done at a general level, asking them to consider their personal futures, the future of their local area and the future of the world. This general approach is well supported by resources in Hicks (2001) and could be developed for key stage 3 geography.

It is possible to envisage developing futures thinking in relation to almost every aspect of geography taught at key stage 3. For example, a futures enquiry could be incorporated into units of work on any of the following:

- The school grounds (e.g. could they be used differently?)
- The local area (e.g. what are the priorities for the future?)
- The local area/town/UK/the world (e.g. what are possible, probable and preferable futures? What are hopes and fears?)
- The European Union (e.g. who wants to join? How might this affect the EU? Should these countries join and under what conditions?)
- The world (e.g. how can we look after our planet for future generations?)
- An area of coastline threatened with erosion (e.g. what are the options? What should we do about it?)
- Climate change (e.g. what might the effects of climate change be on different places?)
- Biodiversity (e.g. what is predicted to happen? Is it important to preserve existing species? Can anything be done about it?)
- Population growth and/or population decline (e.g. what problems are caused by increasing/ declining populations? What can/should be done about it?)
- The town centre (e.g. what would make the town centre a better place?)
- Shopping (e.g. how will we shop in future? What do different groups of people need?)
- Tourism (e.g. how can tourist areas cater for the disabled? What changes are needed?)

- Inequalities in the local area/country/world (e.g. how fair is the present distribution? How could the situation be made fairer?)
- Providing energy for the future (e.g. how are we going to provide for our future energy needs?)
- Recycling (e.g. what are we going to do with all our rubbish?)
- Globalisation and interdependence (e.g. how are we going to relate to the rest of the world?)
- World health, e.g. AIDS (What are the consequences of AIDS in Africa? What are the implications of these consequences? What can we do about it?)

ACTIVITIES

The following activities can be developed to form whole-lesson enquiries, linked with other enquiry work to form a unit of work:

- Consequences
- Forecasting
- Scenarios
- Priorities for the future
- Wild cards
- Values enquiries

CONSEQUENCES

This activity draws on students' existing understanding of cause and effect, on general knowledge and experience and on reasoning ability. It is adapted from Hicks' (2001) use of a futures wheel. The consequences activity tries to answer the question: What are the future implications of this event or situation?

In this activity students are provided with an initial stimulus about an event, a forecast or a possible change. Students consider what the consequences of this would be. The possibilities for the initial stimulus are numerous in geography, including topical local, national and world issues. Some possibilities are:

- the village school is closing down
- a football stadium is being moved to another part of town
- a disused airport has been given the go-ahead to develop as a commercial airport
- Beijing has been selected to host the 2008 Olympic Games
- a new supermarket is being opened
- at the present rate of growth, the world population will double in size in the next 40 years
- air traffic in the UK is expected to increase by 40% in the next five years
- life expectancy in Botswana is projected to be 27 years in the year 2010, rather than the estimated life expectancy of 70 years, which it was projected to be had the AIDS epidemic not struck the country (source: USAid, 2002).

An example of how a consequences activity could be used as part of an enquiry into local services is shown in Example 1. A possible general procedure for using a consequences activity is shown in Example 2. Figure 1 can be used with all the activities in this chapter.

Example 1: Future consequences of changes in services within an area. Source: Rachel Atherton and Nikki Flanagan.

Procedure

Key question:

- What services are there in the local area?
- What impacts would there be in the local area if a service was added or removed?
- How would this affect the future of the area?
- What are the direct and indirect consequences?

Starter activity: Snowballing

Introduction: What types of services are there in your local area? What services do you think are needed in your area?

Students work in pairs to discuss the services in their local area.

- They make a list of services and identify those they consider are most important and those that are missing
- Students join another pair, and in groups of four they discuss the services that are important/missing from their different local areas (students make a new list)

- Each group of four chooses one service from their list that they would like either to add or remove.

Groupwork activity: Consequences

On a piece of flipchart paper, students construct a consequences diagram (Figure 1) to establish the consequences of adding or removing their chosen service.

Plenary: Presentations, discussion and debriefing

Selected groups present their findings from the consequences chart. Possible debriefing prompts: What has your group chosen to add/remove? Why? What are the consequences and why? Which consequences would be short term (1-2 years) and which would be long term (over 10 years)? What do you think about these consequences? What do you think other people in your area would think about these consequences? Do you think these changes are likely to happen? Why or why not?

Figure 1: Framework for consequences activities.

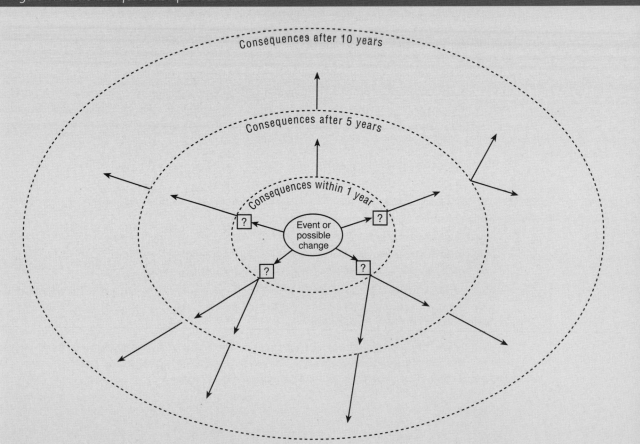

Example 2: Consequences.

Procedure

Key question:

What are the implications of this for the future?

Starter

- Students are presented with information about a startling statistic, an event, a change, or a forecast. This could be in the form of a newspaper headline, a statistic, a proposed change, or a video clip.
- The teacher elicits what students know about the present situation and, if necessary, provides information.
- The teacher encourages students to speculate about what difference this information might make and collects two or three ideas about consequences. These can be marked on a demonstration consequences diagram.

Activity: Consequences

Students work in pairs or small groups to produce a consequences diagram. This is an extended spider diagram, as in Figure 1, on which immediate consequences are noted (first-order consequences, followed by second-order consequences, followed by third-order consequences).

Interim debriefing

This can be done either with the whole class, or with pairs or groups of students. Possible prompts: I wonder what the first consequences you thought of were? Why would it lead to this consequence? Are there any further consequences of this? How did you work that out?

Activity: Categorising consequences

Students work in pairs to categorise their consequences. They could be asked to devise their own way of categorising what they have marked on their diagram. This is more challenging and could lead to better discussion in the debriefing. Alternatively, they could be given a set of categories, such as:

- as likely/unlikely
- short term/long term
- important/less important
- affecting everyone/affecting some people
- positive/negative
- intended/unintended.

Plenary summary debriefing

Possible debriefing prompts: I wonder how you have categorised the consequences? Which have you identified as (for example) more important? Why are they more important? What do you think of your predicted consequences? Is it what you would like to happen in the future? Could we do anything to help it to happen or prevent it from happening?

FORECASTING

The activity of forecasting the future is based on the commonly used method of trend analysis to answer the question 'What is the probable future?' Trend analyses can be used for any topics for which statistical data for the recent past is available. It is often used for futures thinking about population. Past trends are projected into the future to produce tentative 'forecasts', not 'predictions'. These forecasts become starting points for thinking about the future. The activity of forecasting the future could be used in key stage 3 geography, in investigations of:

- population growth in the world, or in a particular country, e.g. China
- the decline of population, e.g. Italy
- population data such as birth rates, death rates, life expectancy rates
- changes in climate
- decline in natural resources or wildlife, e.g. forests, fish stocks, elephants, black rhinoceros
- increase in consumption of resources.

Information about population can be found at websites for the United Nations, for the Population Reference Bureau, and for the US Census Bureau. The latter website gives access to population pyramids for 2000, 2025 and 2050 and dynamic changing population pyramids for every country. Information about natural resources can be found at the Worldwide Fund for Nature (WWF) website (www.wwf-uk.org).

Clearly, if trend analysis is used in geography at key stage 3, forecasts are likely to be estimates based on visual extension of trends evident in graphs, rather than on the kinds of mathematical models used in organisations and government. However, whether the forecasts are rough estimates or the result of sophisticated calculations, it needs to be recognised that trend analysis is based on what happened in the past and that future trends could be different. Forecasts can never be certain. An enquiry procedure is shown in Example 3.

Example 3: Forecasting.

Procedure

Key question

What will happen in the future?

Starter

The teacher elicits what the students already know about the situation being studied. Intelligent guesswork is used to invite students to guess the current figures.

The teacher presents a scenario in which there is a need to know about what will happen in the future, e.g. future population of world or country, for a television programme report.

Activity: Forecasting the future

- Students draw or study a graph of recent trends related to issue being investigated.
- Students estimate what will happen in the future and mark it on the graph.

Interim debrief

For example: What does the graph tell you about the situation at the moment? How did you work out how to extend the graph? What does your graph tell you about the situation in 5 (10 or 50) years time?

Activity: Consequences

Students work in groups to draw a consequences diagram (Figure 1) to show what the consequences would be if this trend continued.

Plenary summary debrief

Possible debriefing prompts: I wonder how likely your forecasts are? What do you think? What might stop it happening? What would the consequences of that be? Do you want your forecasts to become true?

Additional activity: Wild card

- A wild card is chosen
- Students discuss how it would affect their graph and draw a new forecast
- Discussion of the impact of the wild-card event is included in the plenary debrief.

SCENARIOS

Scenarios are stories about possible futures. They are used widely in futures thinking in organisations and governments to explore different possible futures. Scenarios need to be different from the present in some way and they have to be possible.

Hicks (2001) provides an example of future scenarios in *Citizenship for the Future*. He presents annotated illustrations of four alternative future environments:

- more of the same (continuing present trends)
- technological fix (emphasising possibilities provided by technology)
- edge of disaster (emphasising worst futures)
- sustainable development (emphasising conservation of resources and preservation of environmental quality).

Hicks presents an accompanying activity for students in which they are asked to:

- study a particular scenario
- identify good things
- identify some of the difficult things
- decide who will benefit and who will lose in this future
- decide whether they would like to live in this future.

This activity can be used either with the excellent illustrations in *Citizenship for the Future* (Hicks, 2001) or teachers can devise and present three or four futures scenarios for the topic they are studying.

A possible procedure is set out in Example 4.

Example 4: Scenarios. Source: Rachel Atherton and Nikki Flanagan.

Procedure

Key questions

- What are the possible futures?
- What do we think of them?

Starter

- Elicitation of initial ideas about the types of scenarios to be introduced (e.g. What changes seem likely if present trends continue? What would a technologically advanced future bring?)
- Organise students into groups. The number of people in each group should be the same as the number of scenarios being discussed.

Individual activity: Scenarios

- Within the group, each student is provided with a different scenario to study individually (e.g. technological fix, more of the same, edge of disaster, sustainable development).
- Students individually identify the good/bad points about their particular scenario.

Groupwork activity: Scenarios

- Each student reports to their small group about the future scenario they have studied.
- The group decides collectively which future they would prefer to live in and why.

Plenary: Sharing ideas and debriefing

- Sharing what the group has decided: make a tally of each group's decision to find out which is the preferred future of the class.
- Possible debriefing prompts: Which groups chose the most preferred future? Why did they choose this? What did they like about it? Was there anything they did not like about it? Did other groups choose other futures? Why? What kinds of disagreements were there in the groups? How were they resolved? How did they reach their decision? How might we make our preferred futures happen?

Follow up/homework

- Students write about their preferred future, stating what they prefer and why they prefer this.

PRIORITIES FOR THE FUTURE

Many organisations and local authorities try to involve the general public in determining the priorities for the future. These might be investigated through a survey or through public meetings or futures workshops. They are applied to situations where there are many possibilities but limited funding.

Example 5a is an activity based on a survey carried out during 2002 by Sheffield First Partnership to find out what people thought would make Sheffield a more successful city. Sheffield residents were invited to comment on specified 'features of a successful city' (Example 5b). This could easily be modified to be appropriate for other towns and cities.

Example 5 : Priorities for the future.

(a) Procedure

Key question

What will make Sheffield a more successful city?

Starter

- Sheffield First Partnership wants to find out what priorities people living there have for the future of the city. The aim of this lesson is to find out views about the future of Sheffield and to write a letter about this.
- The task is to think carefully about the features on the cards (Example 5b) and to decide what is important and what is not important.

Activity

- Students (in pairs) look through the set of cards.
- They sort the cards into the following categories according to whether they think the feature is:
  1. important for Sheffield
  2. important for themselves and their family
  3. important both for Sheffield and for themselves and their family
  4. not important.
- Students are invited to consider whether there are other things that are important to them that are not listed on the cards. They write these on the blank cards.
- They make a note of how they have categorised the cards.

Alternative activity

- Students (again in pairs) look through the set of cards.
- They discard one card.
- They decide whether they want to replace one of the remaining cards with a card of their own choice.

- They arrange the nine cards in a diamond ranking pattern (see Example 3, Chapter 10, pages 136-137).

Activity in groups of four

- Each pair joins another pair and they compare the way they have categorised the cards.
- The students discuss why they think each feature is important or unimportant.
- The group decides which three features are the most important (including any features added on the blank cards).

Plenary discussion

Students share information about categories to find out whether there are any features everyone agrees on and those about which there are disagreements. These are listed for all to see. A tally is made of the three most important features chosen by each group. Debrief: What were your reasons for thinking that some features were important? What were your reasons for thinking that some features were unimportant? Why are there differences of opinion? (if there are).

Extended writing

- Preparation for writing a letter to Sheffield First Partnership about three top priorities. Whole-class discussion: How would they start the letter? Ask each group to draft a short paragraph about one priority stating what it is and why it is important. Share some of these paragraphs. Invite comments. How would they finish the letter?
- Students write letters which will be sent to Sheffield First Partnership.

Example 5 : Priorities for the future ... continued

(b) Priorities for the future: features of a successful city

Low crime A city where people feel safe to live and work	Attractive neighbourhoods Where each neighbourhood is a pleasant place to live and visit, where people feel proud of their area	Good transport People can get around the city quickly, easily and cheaply and there are good road and rail links to other cities and international airports
A strong economy A modern economy with good opportunities for the whole community and businesses that make the region strong again	Good education Good schools, colleges and universities, and more learning opportunities for people to achieve their full potential	Good health A healthy place to live, where people have good local health services
A lively city centre A welcoming city centre with excellent shopping, leisure and cultural facilities	Good cultural facilities Sporting events, theatres, galleries and museums, live music concerts, clubs and festivals which are popular with both local people and visitors to the city	Good cultural facilities A city that welcomes people of every background and where new residents contribute to the city
Well-regarded A city where others would like to live, visit or do business	Other	Other

WILD CARDS

Wild cards suggest events which may be unlikely to happen but which would have a big impact if they did. In industry and government, wild-card methods are mainly concerned with negative events and risk. It would be unfortunate to use only negative wild cards in the classroom. Views of people in the UK about the future are generally pessimistic (Hicks and Holden, 1995) and such cards could enhance fears and distort futures thinking. It would be preferable to use both positive and negative wild cards in futures thinking or cards of events that would be positive for some people and negative for others. Clearly, the types of events described on the wild cards will depend on what is being studied. For example, there could be wild cards related to the following:

- additional funding for a new project in an area, e.g. piped water supply to a village in a LEDC or lottery funding for a theme park in the UK
- a policy decision to use green belt land for housing
- pollution of a large area of coastline.

This activity is best introduced half-way through the other activities suggested above, to stimulate creativity and fresh thinking. The procedure is simple:

- a card is chosen from a small pack of wild cards, created by the teacher
- the event on the card is read out
- students discuss what the consequences of this event would be
- ideas are shared.

The activity could be used with each group choosing a different wild card, discussing the implications and presenting their event and possible consequences to the whole class. Alternatively, one wild card could be chosen for the whole class, pairs or groups of students could discuss the implications, and then these ideas could be shared and compared in whole-class debriefing.

FUTURES ENQUIRIES AND VALUES ENQUIRIES

Many values enquiries (Chapter 11, pages 141-151), using activities such as role play, are concerned with decisions about the future. These activities can make an important contribution to developing futures thinking.

INTRODUCTION

This chapter is based on enquiry work developed by Steve Wilson, Head of Geography at Great Baddow High School, Chelmsford in Essex. The school is a mixed comprehensive school with a wide range of levels of achievement. Every year, between November and the summer half term, year 8 students investigate the natural hazards which take place during that period. This enquiry work, planned as part of the geography course, is on a totally different scale from all the other enquiry activities presented in this book: it is a really big project, in the time allowed for the work, in what is expected from the students and in its impact. It is an open-ended project demanding independent research from students.

LEARNING OPPORTUNITIES

'Sophie could remember situations when her mother or the teachers at school had tried to teach her something she had not been receptive to. And whenever she had really learned something, it was when she had somehow contributed to it herself'

(Gaardner, 1995, p. 47).

During the course of the big project students have opportunities to:

- study topical events
- increase their factual knowledge about natural hazards
- develop understanding of geographical processes
- recognise and explain distribution patterns
- recognise development issues
- empathise with the plight of people
- develop a wide range of skills including:
  - o searching for information
  - o selecting information
  - o using atlases
  - o mapping
  - o ICT skills
  - o literacy
  - o numeracy
  - o presentation and organisation.

WHERE DOES THE BIG PROJECT FIT IN?

The idea of the 'big project' completed over a period of months is particularly well-suited to the study of natural hazards and the widespread media coverage they attract. The project described in this chapter could be adapted for other aspects of the key stage 3 geography national curriculum, including:

- Changes in the local area
- Weather and climate (UK, Europe, or world)
- The geography of sport

HOW THE NATURAL DISASTERS ENQUIRY IS CARRIED OUT?

The project is developed in seven stages, but it can be further extended – each of the stages is outlined below.

Stage 1: Introduction to the project (one lesson)

- Students are given an information and task sheet (Example 1).
- Students are introduced to the key questions which frame the enquiry: What natural disasters

are there? (During the period of study.) When have they occurred? Where have they occurred? What effect have they had?

- The meaning of 'natural hazard' is discussed and distinguished from disasters which are mainly the result of human action.
- Possible ways of organising the information are described and discussed. Students can organise the material either into a continuous diary of events or into categories of hazards. (Lower achieving students tend to opt for the continuous diary arranged in chronological order. Higher achieving students are encouraged to categorise the natural hazards and arrange records of each type of hazard in chronological order.)
- Attention is drawn to the way the marks will be awarded for the big project.
- Students are provided with outline maps of the world on which they can plot events as they occur.

Example 1: Information and task sheet: natural hazards.

Year 8 natural hazards project (November–May)

A natural hazard is something that occurs due to a natural event (not caused by humans), which causes problems (or possible problems) for people and/or property.

A list of hazards might include:

- Earthquakes
- Landslides
- Drought
- Fog (bad visibility)
- Snow and ice (on roads)
- Heat (heat stroke)
- Strong wind (e.g. hurricanes, tornadoes, gales)
- Volcanoes
- Floods
- Avalanches
- Frost (on plants/roads/pipes)
- Hail (crop damage)
- Cold (hypothermia)

Task

Between now and the first week back after the May half-term, when the project must be handed in – you must collect information about natural hazards and present the work in a *scrapbook* or a *ring binder*.

Information should be collected from:

- Television (including Teletext or CEEFAX)
- Newspapers (national and local)
- Internet
- Radio
- Magazines
- Own experience

You must collect the information in either *diary* form or under *headings* for each kind of hazard.

Always say *when* and *where* each event happens. Try to discover what the hazard was like and what effects it had on the landscape and people.

Try to use a mixture of your own writing and articles and pictures that can be neatly cut out and stuck in.

You will also be given a *world map* on which to mark where the events occur.

You may add some general background information about particular types of hazard, but remember the emphasis should be on *current events* and most of the marks will be awarded for your careful observations about what is going on in the world between now and May.

Marks will be awarded as follows:	
Content	10
Effort	5
Presentation	5
Total	20

A bronze award will be given for all scores of 16 or more.

Scores of 19 and 20 will receive a special certificate.

All work will be checked during the first two weeks after Christmas to see how you are getting on. There will be an allowance of a homework every third week to help you complete the work, but it will probably take a bit longer than this and you are advised to spend time on it whenever a hazard occurs.

When you have finished your project you must evaluate your work, commenting on what you have learned and how interesting you found collecting and organising the information.

Get your family involved, as they might be able to help you.

Good luck with your project!

- Students often ask, 'How much do we have to write?' to which they receive the reply, 'It depends on what happens'. In practice, higher achieving students who are industrious find huge amounts of detail. In 2002, one student submitted 200 A4 pages of information!

Stage 2: Independent research

- Students carry out the research at home, using their own time and allocated homework time (one homework every three weeks).
- Students often involve their parents and other relations in the project. Some have contacted pen friends abroad in order to collect information.

Stage 3: Interim monitoring and discussion

- All work is checked in the first two weeks after the Christmas holiday.
- When natural disasters happen they are brought to the attention of the class and occasionally lesson time is used to discuss them.

Stage 4: Independent research

- Students continue to carry out their research.

Stage 5: Evaluation of the project

- Students are expected to evaluate their work by commenting on what they have learned.

Stage 6: Follow-up lesson

- Students share their findings with other students.

Stage 7: Extending the audience for the project

The audience for the project is much larger than that of teacher as assessor of the work. It can include:

- Other members of the class
- Senior members of staff who are shown some projects
- Parents and other visitors to the school on open evenings.

Going further

Sometimes groups of students have responded to a major disaster during the period of study by organising a collection of money to give to the disaster relief fund - an example of citizenship in action.

ASSESSMENT

The big project is assessed on content, effort and presentation with a possible total of 20 marks (Example 1).

- **Content:** To receive the highest marks for content, students must include a range of hazards, over a range of time. They must include background information and include their own views in their evaluation.
- **Effort:** The effort grade depends on how much time students have spent on the project and how the work relates to what can be expected from them (related to which set they are in).
- **Presentation:** This includes neatness, how well-organised the information is, whether the events described are dated, the variety of materials used, the use of titles, subtitles and innovative illustrations.

Students with over 16 marks are given a bronze award, while those with 19 or 20 marks receive a special certificate.

WHAT STUDENTS SAY ABOUT THE PROJECT

The students' evaluations of their work are an important part of the big project. Example 2 provides evidence of the value of the work to one student. It increased this student's knowledge and understanding of natural hazards, developed a wide range of enquiry and ICT skills, and promoted collaborative working with other members of the family.

Example 2: Extracts from students' evaluations of the big project on natural hazards.

Misty

In this project, I have learned a lot about the world, what happens to different countries, how much damage natural hazards make. Doing this project has challenged me, trying to find hazards on the Internet and the use of newspapers for weather reports. I have expanded the use of my computer skills, by word-processing my work, but also accessing the internet to find the appropriate information for this project.

Kathryn

Over the past six months, I have collected many hazards for my project, many, when I look back, cause terrible damages to the places. Nearly all the entries of the hazards I have, have reported people dead, or seriously injured. Reading back, I cannot believe how many natural hazards take place every day. This project has also helped me with my map skills.

Sam

I have enjoyed working on this project, as I find it very informative, and interesting. My whole family has participated in this project, by finding useful information from their newspapers!!

Edward

I think I am fortunate that I live in Great Britain as we rarely have extreme weather conditions such as floods and heavy snows, other countries experience not only these but earthquakes, volcanic activity and hurricanes!!

Extracts from the evaluations of a range of students provides evidence that the big project was valuable to them in similar ways. All students increased their knowledge and understanding:

- 'I found out where many countries are.'
- 'From doing this project I have learnt all about different types of hazards and that they occur all around the world.'
- 'I have learnt that something doesn't have to kill anyone to be a hazard.'
- 'Disasters have a great effect on people's lives in different ways, from loss of their home and belongings, to the loss of life, all of which are equally devastating.'
- 'There are a lot more natural disasters than I ever knew there were, all happening around the same time. On the other hand, I found that they can just have little effect.'

The big project increased their awareness of the world and topical events:

- 'It has also helped me to open my eyes to the world around me.'
- 'It has made me think more about the world.'
- 'If I hadn't been doing it I wouldn't have been aware of half the disasters that are happening every day all over the world.'

Many of the evaluations demonstrated how valuable parental support could be:

- 'I did this project mostly on my own, but with the help of my parents telling me how to make it better, catch the reader's eye, etc. Friends helped me on the Internet, etc., to obtain more information.'
- 'The people that helped me with the project are my mum and dad. They kept pointing things out and kept loads of papers.'

Additional evidence of family involvement has been provided at parents' evenings where parents frequently commented on the big project. One parent commented on how the project had given the family something to work on and talk about together. The evaluations showed the personal impact the project made on the students. They were affected by what they saw and learnt and related it to their own lives:

- 'The pictures I have seen when searching for information to use in my project are amazing.'
- 'I found that it hits the poorest countries especially the volcano in Goma where children were climbing over really hot molten lave to carry anything they can find whether it belongs to them or not. The picture of the boy carrying a piece of metal for a makeshift home and the boy making his meal in a pan on the hot rocks made me think how lucky I am.'

The evaluations provided an example of what students' 'ownership' of the work might mean; they wrote about 'my project' and even 'my earthquakes':

- 'This project has helped me learn about the hot spots for hazards, to help explain this the Philippines are to me a hot spot for earthquakes as nearly all my earthquakes have happened around that area.'

There was general approval for the big project, in spite of the hard work involved:

- 'I found this project very good.'
- 'Generally, this project has been fun and interesting.'
- 'I found it very hard work and had to give up loads of time in the holidays to do it as there are so many hazards happening all over the world.'

SUMMARY

This big project on natural hazards gives students the opportunity to get involved in independent research over a long period of time, collecting and organising their own data. It gives them the opportunity to develop a wide range of geographical enquiry skills as well as a range of key skills such as numeracy, literacy, the use of ICT and communication skills. Through the use of these skills and through reflecting on their learning they increase their knowledge and understanding of natural hazards.

This project achieves important geographical objectives, but it also has wider significance. It encourages students to recognise that they can learn about the world, not only in the geography classroom, but also outside the classroom, through the media and through talking to other people. If students are to continue to increase their geographical understanding after they have stopped studying geography as a subject, then it is important to enable them to learn from the world outside the classroom and to value this learning. Furthermore, the project benefits from students collaborating with parents and relations. Most parents are only too keen to support their children's learning. The big project provides a vehicle for them to do so.

If enquiry work is limited to single-lesson enquiries, however well these are planned and carried out students will not have the opportunity to develop the kind of ownership of their work that has been demonstrated in the natural hazards project. There are strong arguments for incorporating extended enquiries into the key stage 3 curriculum, and for giving students responsibility for their own research. The natural hazards enquiry shows an excellent way of doing this.

REFERENCES

Abbs, P. and Richardson, R. (1990) *The Forms of Poetry.* Cambridge: Cambridge University Press.

Allen, J. and Massey, D. (1995) *Geographical Worlds.* Milton Keynes: Open University Press.

Ball, S.J. and Bowe, R. (1992) 'Subject departments and the "implementation" of national curriculum policy: an overview of the issues', *Journal of Curriculum Studies,* 24, 2, pp. 97-115.

Barnes, D. (1976) *From Communication to Curriculum.* Harmondsworth: Penguin Books.

Barnes, D. (1982) *Practical Curriculum Study.* London: Routledge and Kegan Paul.

Barnes, D. and Todd, F. (1995) *Communication and Learning Revisited.* Portsmouth, USA: Boynton/Cook Publishers Inc.

Barnes, D., Johnson, G., Jordan, S., Layton, D., Medway, P. and Yeoman, D. (1987) *The TVEI Curriculum 14-16: An interim report based on case studies in twelve schools.* University of Leeds.

BBC (1992) *Pole to Pole with Michael Palin* (video). London: BBC Enterprises Ltd.

Birmingham DEC (1995) *The Compass Rose.* Birmingham: Development Education Centre.

Black, P. and Wiliam, D. (1999) *Assessment for Learning: Beyond the black box.* Cambridge: University of Cambridge School of Education.

Bloom, B.S. (1956) *Taxonomy of Educational Objectives.* New York: David McKay.

Bright, N. and Leat, D. (2000) 'Towards a new professionalism' in Kent, A. (ed) *Reflective Practice in Geography Teaching.* London: Paul Chapman Publishing, pp. 253-61.

Britton, J., Burgess, T., Martin, N., McLeod, A. and Rosen, H. (1976) *The Development of Writing Abilities (11-18).* London: Macmillan.

Bruner, J. (1966) *Toward a Theory of Instruction.* New York: W.W. Norton and Company Inc.

Bruner, J. (1986) *Actual Minds, Possible Worlds.* Cambridge MA: Harvard University Press.

Bruner, J. (1996) *The Culture of Education.* Cambridge MA: Harvard University Press.

Butt, G. (1993) 'The effects of audience-centred teaching on children's writing in geography', *International Research in Geographical and Environmental Education,* 2, 1, pp. 11-24.

Butt, G. (2001) *Theory into Practice: Extending writing skills.* Sheffield: Geographical Association.

Cairney, T.H. (1995) *Pathways to Literacy.* London: Cassell.

Carroll, L. (1998) *Alice's Adventures in Wonderland and Through the Looking Glass.* Oxford: Oxford University Press.

Carter, R. (ed) (1991) *Talking About Geography: The work of geography teachers in the National Oracy Project.* Sheffield: Geographical Association.

Cattaneo, P. (director) (1997) *The Full Monty.* Twentieth Century Fox Film Corporation.

Christian Aid (2001) *The Trading Game.* London: Christian Aid.

Cooper, H. (1992) *The Teaching of History.* London: David Fulton.

Counsell, C. (2001) 'Challenges facing the literacy co-ordinator' in Strong, J. (ed) *Literacy Across the Curriculum: Making it happen.* London: Collins Educational, pp. 14-15.

D'Arcy, P. (1989) *Making Sense, Shaping Meaning.* Portsmouth NH: Boynton Cook/Heinemann.

D'Arcy, P. (2000) *Two Contrasting Paradigms for the Teaching and Assessment of Writing.* London: National Association for the Teaching of English.

Dalton, T.D. (1988) *The Challenge of Curriculum Innovation: A study of ideology and practice.* Lewes: Falmer.

Davidson, G. and Catling, S. (2000) 'Toward the question-led curriculum 5-14' in Fisher, C. and Binns, T. (eds) *Issues in Geography Teaching.* London: Routledge/Falmer, pp. 271-95.

de Saint-Exupery, A. (1942) *The Little Prince* (first translated into English, 1945, new translation by Wakeman, A. (1997)). London: Pavilion Classics.

REFERENCES

DES (1975) *A Language for Life*. London: HMSO.

DES (1995) *Geography in the National Curriculum (England)*. London: HMSO.

Dewey, J. (1933) *How We Think: A re-statement of reflective thinking upon educational processes*. Boston: Heath.

DfEE (1999a) *Geography: The national curriculum for England*. London: DfEE.

DfEE (1999b) *The National Curriculum: Handbook for secondary teachers in England (KS3&4)*. London: DfEE.

DfEE (1999c) *Mathematics: The national curriculum for England*. DfEE

DfEE (2001a) *Key Stage 3 National Strategy: Literacy across the curriculum*. London: DfEE.

DfEE (2001b) *Key Stage 3 National Strategy: Framework for teaching mathematics – years 7, 8 and 9*. London: DfEE.

DfES (2001) *Key Stage 3 National Strategy. Numeracy across the curriculum – Notes for school-based training*. London: DfES.

Driver, R., Squires, A., Rushworth, P. and Wood-Robinson, V. (1994) *Making Sense of Secondary Science: Research into children's ideas*. London: Routledge.

Eliot, T.S. (1943) *The Four Quartets*. London: Faber & Faber.

Gaardner, J. (1995) *Sophie's World*. London: Phoenix House.

Ghaye, A. and Robinson, E. (1989) 'Concept maps and children's thinking: a constructivist approach' in Slater, F. (ed) *Language and Learning in the Teaching of Geography*. London: Routledge, pp. 115-48.

Goodey, B. (1971) *Perceptions of the Environment*. Birmingham: Centre of Urban and Regional Studies, University of Birmingham.

Greig, D. (2000) 'Making sense of the world: language and learning in geography' in Lewis, M. and Wray, D. (eds) *Literacy in the Secondary School*. London: David Fulton, pp. 69-90.

Hicks, D. (2001) *Citizenship for the Future: A practical classroom guide*. Godalming: WWF-UK.

Hicks, D. and Holden, C. (1995) *Visions of the Future*. Chester: Trentham Books.

Hopkirk, G. (1998) 'Challenging images of the developing world using slide photographs', *Teaching Geography*, 23, 1, pp. 34-5.

Jackson, P. (1987) *Maps of Meaning*. London: Routledge.

Jackson, P. (2000) 'New directions in human geography' in Kent, A. (ed) *Reflective Practice in Geography Teaching*. London: Paul Chapman Publishing, pp. 50-6.

Jackson, P. (2002) 'Geographies of diversity and difference', *Geography*, 87, 4, pp. 316-23.

Jones, P. (1981) 'Turning the beer brown'. Trowbridge: Wiltshire Learning about Learning booklet number 11 (unpublished).

Laing, R.D. (1971) *Knots*. Harmondsworth: Penguin Books (and poem quoted at www.oxforddntc.co.uk/'knots'.htm).

Lambert, D. and Balderstone, D. (2000) *Learning to Teach Geography in the Secondary School: A companion to school experience*. London: Routledge/Falmer.

Leat, D. (ed) (1998) *Thinking Through Geography*. Cambridge: Chris Kington Publishing.

Leat, D. and Chandler, S. (1996) 'Using concept mapping in geography teaching', *Teaching Geography*, 21, 3, pp.108-12.

Leat, D. and Kinninment, D. (2000) 'Learn to debrief' in Fisher, C. and Binns, T. (eds) *Issues in Geography Teaching*. London: Routledge, pp. 152-72.

Leat, D. and Nichols, A. (1999) *Theory Into Practice: Mysteries make you think*. Sheffield: Geographical Association.

Lewis, M. and Wray, D. (1995) *Developing Children's Non-fiction Writing.* Leamington Spa: Scholastic.

Lunzer, E. and Gardner, K. (1979) *The Effective Use of Reading.* Oxford: Heinemann.

Massey, D. (1998) 'The spatial construction of youth cultures' in Skelton, T. and Valentine, G. (eds) *Cool Places.* London: Routledge, pp. 121-9.

Matthews, H. (1992) *Making Sense of Place: Children's understanding of large-scale environments.* Hemel Hempstead: Harvester Wheatsheaf.

Matthews, H., Limb, M. and Taylor, M. (2000) 'The "street as thirdspace"' in Holloway, S.L. and Valentine, G. (eds) *Children's Geographies: Playing, living and learning.* London: Routledge, pp. 63-79.

Mercer, N. (1995) *The Guided Construction of Knowledge.* Clevedon: Multilingual Matters Ltd.

Mercer, N. (2000) *Words and Minds.* London: Routledge.

Naish, M., Rawling, E. and Hart, C. (1987) *Geography 16-19: The contribution of a curriculum project to 16-19 education.* London: Longman.

Neighbour, B.M. (1992) 'Enhancing geographical inquiry and learning', *International Research in Geographical and Environmental Education,* 1, 1, pp. 14-23.

Nichols, A. with Kinninment, D. (2001) *More Thinking Through Geography.* Cambridge: Chris Kington Publishing.

Norman, K. (ed) (1992) *Thinking Voices.* London: Hodder and Stoughton.

Ofsted (2002) downloadable from website www.ofsted.gov.uk/reports/107/107140.pdf

Palin, M. (1997) *Full Circle.* London: BBC Publications.

Parsons, C. (1987) *The Curriculum Change Game.* London: Falmer.

PIU (2001) *A Futurist's Toolbox: Methodologies in futures work.* London: The Stationery Office.

Pomeroy, J. (1991) 'The press conference: a way of using visiting speakers effectively', *Teaching Geography,* 16, 2, pp. 56-8.

Pool, H., Brockes, E. and Phipps, C. (1999) 'Who do we think we are?', *The Guardian,* 20 January (available on website: http://www.guardian.co.uk/g2/story/0,3604,323031,00.html).

QCA (1998) *Geographical Enquiry at Key Stages 1-3: A discussion paper.* London: QCA.

QCA/DfEE (2000) *Geography: A scheme of work for key stage 3.* London: QCA.

Raths, J. (1997) 'Identifying activities that seem to have inherent worth', *Prospero,* 3, 2, pp. 67-8.

Rawling, E. (2001) *Changing the Subject: The impact of national policy on school geography 1980-2000.* Sheffield: Geographical Association.

Reid, J., Forrestal, P. and Cook, J. (1989) *Small Group Learning in the Classroom.* Scarborough: Chalk Face Press.

Relph, E. (1976) *Place and Placelessness.* London: Pion.

Riley, C. (1997) 'Evidential understanding, period knowledge and the development of literacy: a practical approach to "layers of inference" for key stage 3', *Teaching History,* pp. 6-10.

Riley, M. (1999) 'Into the key stage 3 history garden: choosing and planting your enquiry questions', *Teaching History,* pp. 8-13.

Roberts, M. (1995) 'Interpretations of the geography national curriculum: a common curriculum for all?', *Journal of Curriculum Studies,* 27, 2, pp. 187-205.

Roberts, M. (1996) 'Teaching styles and strategies' in Kent, A., Lambert, D., Naish, M. and Slater, F. (eds) *Geography in Education.* Cambridge: Cambridge University Press, pp. 231-59.

Roberts, M. (1998a) 'The nature of geographical enquiry at key stage 3', *Teaching Geography,* 23, 4, pp. 164-7.

Roberts, M. (1998b) 'The impact and legacy of the 1991 geography national curriculum at key stage 3', *Geography,* 83, 1, pp. 15-27.

Rosen, M. (1994) 'School students' writing: some principles' in Brindley, S. (ed) *Teaching English.* London: Routledge, pp. 195-201.

Sayer, A. (1985) 'Realism and geography' in Johnson, R.J. (ed) *The Future of Geography.* London: Methuen, pp. 159-73.

SCAA (1994) *The National Curriculum and its Assessment.* London: SCAA.

Schools Council (1980) *Man, Land and Leisure: Teachers' guide.* Walton-on-Thames: Thomas Nelson and Sons Ltd.

Sheeran, Y. and Barnes, D. (1991) *School Writing: Discovering the ground rules.* Buckingham: Open University Press.

Shurmer-Smith, P. (2002) *Doing Cultural Geography.* London: Sage

Smith, D. (1977) *Human Geography: A welfare approach.* London: Edward Arnold.

Spooner, D. (2001) 'Odysseys in a regional world', *Geography,* 86, 4, pp. 287-303.

Stenhouse, L. (1975) *An Introduction to Curriculum Research and Development.* London: Heinemann.

Tolley, H. and Reynolds, J.B. (1977) *Geography 14-18: A handbook for school-based curriculum development.* London: Macmillan.

Traves, P. (1994) 'Reading' in Brindley, S. (ed) *Teaching English.* London: Routledge, pp. 91-7.

USAid (2002) 'Life expectancy will drop worldwide due to Aids', Press release 8 July (www.usaid.gov/press/releases/2002/pr020708.html).

Valentine, G. (2001) *Social Geographies: Space and society.* Harlow: Prentice Hall.

Vygotsky, L.S. (1962) *Thought and Language.* Cambridge MA: the Massachusetts Institute of Technology Press.

Webster, A., Beverldge, M. and Reed, M. (1996) *Managing the Literacy Curriculum.* London: Routledge.

Wheen, F. (2002) *Hoo-hahs and Passing Frenzies: Collected journalism.* London: Atlantic Books.

Wood, D., Bruner, J. and Ross, G. (1976) 'The role of tutoring in problem solving', *Journal of Child Psychology and Psychiatry,* 17, 2, pp. 89-100.

Wray, D. and Lewis, M. (1997) *Extending Literacy: Children reading and writing non- fiction.* London: Routledge.

Zephaniah, B. (2001) *Refugee Boy.* London: Bloomsbury Publishing.

Websites

The websites listed below are all mentioned in the text. These uniform resource locators were correct at the time of going to press, but please bear in mind that websites and access to them changes frequently.

- BBC News – http://news2.thls.bbc.co.uk/hi/english/world/europe/newsid%5F561000/561754.stm (information on French oil spill)
- Child Hope UK – www.childhopeuk.org/voices (life of street children in Nairobi, Kenya)
- Edinburgh University Geography Department – www.geo.ed.ac.uk/quakes/quakes.html (worldwide earthquakes locator)
- Google – www.google.co.uk (images section is an excellent source of images of places and geographical themes)

- GreenPeace – www.greenpeace.org (environmental perspectives on oil)
- Human emotions – www.human-emotions.com (exploring feelings – 412 separate feelings arranged in 24 groups)
- Infoplease.com – www.infoplease.com/ipa/A0778562.html (information on development and country rankings)
- Kenyan newspaper article – www.mg.co.za/mg/news/98apr1/15apr-kenya.html (life of street children in Nairobi)
- Oil information – http://seawifs.gsfc.nasa.gov/OCEAN_PLANET/HTML/peril_oil_pollution.html (oil spills)
- Petroleum – www.petroleum.co.uk/politics/htm (excellent site for information on oil reserves)
- Population Reference Bureau – www.prb.org/ (US population information)
- Probe international – www.probeinternational.org (Three Gorges Dam construction)
- Southampton University – www.soton.ac.uk/~engenvir/environment/water/oil.slicks.html (information on oil spills)
- Swansea University – www.swan.ac.uk/biosci/empress/ (Sea Empress oil spill)
- UK Department of Trade and Industry – www.dti.gov.uk/EPA/eib/ (government information on energy)
- United Nations Development Programme – www.undp.org/hdr2002 (information on development and country rankings)
- United Nations Population Information Network – www.un.org/popin/ (worldwide population information)
- US Census Bureau – www.census.gov/ (population information for every country)
- Worldwide Fund for Nature (WWF) – www.wwf-uk.org (information about natural resources)